Tamed & Untamed

Praise for

"In their writing and in their lives and in their remarkable friendship, Liz and Sy break down false barriers and carry us closer to our fellow creatures."

—from the foreword by VICKI CONSTANTINE CROKE,
author of *Elephant Company*

"Two kindred spirits treat animals as they ought to be treated, with understanding, knowledge, and humor. These well-crafted essays are a pleasure to read and make you marvel at our fellow travelers on this planet."

—FRANS DE WAAL, author of
Are We Smart Enough to Know How Smart Animals Are?

"Are humans the ultimate species? Nope, not according to these authors. Water bears are: They have been around for 500 million years and will survive after we destroy the planet. Will chickens in your backyard sooner or later give you a name? Do eels dream? Can an octopus have a sense of humor? Read and learn. This is an absolutely enchanting book that anybody who loves animals will not only want to own but also to give to good friends. It is full of precious lines and deep wisdom, and there is a delicious sense of humor throughout. The authors constantly bring us back to the recognition that we are just one among millions of remarkable animals, each one worthy of study and deep admiration, the kind these very authors accord them."

—JEFFREY MOUSSAIEFF MASSON, author of
Dogs Never Lie About Love; coauthor of *When Elephants Weep*

"Two of the most tuned-in people in the world have now given us these marvelous narratives of nonhuman beings living their lives on our shared planet. This is exactly what we need more of. We need to understand *who* we are here with. And, more and more urgently, to understand that we are not alone on our planet. As humans become more isolated and alienated, stories of other animals offer us our best chance for succeeding at being human."

—CARL SAFINA, author of
Beyond Words: What Animals Think and Feel

"I cannot recommend highly enough this memorable collection of essays about the secret life of animals from two of the most thought-provoking, animal-savvy writers of this time, Sy Montgomery and Elizabeth Marshall Thomas. A must-read for anyone interested in all creatures great and small."

—NICHOLAS DODMAN, DVM, author of *Pets on the Couch*; professor emeritus, Tufts University

"A beautifully written, fascinating compendium of essays about many species with whom we share our planet. Sy Montgomery and Elizabeth Marshall Thomas impart their insights into the behavior of a variety of nonhumans in this extremely informative and thought-provoking read."

—IRENE M. PEPPERBERG, author of *Alex & Me*

"*Tamed and Untamed* is a beautiful duet between two of the world's finest nature writers. These enchanting essays bring to life creatures both novel and familiar, from pink dolphins to domestic dogs, war elephants to garden slugs. Each chapter reveals a new animal mystery and adds to the menagerie of our minds."

—ABIGAIL TUCKER, author of *The Lion in the Living Room*

"*Tamed and Untamed* is a gem of a book. Written by two incredibly gifted writers, it's a multicourse buffet of wonderful and thought-provoking stories about the surprising and wide-ranging intelligence, and deep and rich emotional lives of many different nonhuman animals. These eloquent authors weave solid science into their stories so that nonresearchers can well understand what is happening in the heads and hearts of the dogs, cats, rats, hawks, octopuses, and many other animals about whom they write. The bottom line is that we are not all that unique among the fascinating and diverse beings who are called 'animals' and with whom we share our magnificent planet. They, like this book, are gifts we must cherish."

—MARC BEKOFF, author of *Rewilding Our Hearts*; coauthor of *The Animals' Agenda*

"Who but Sy Montgomery could describe a hawk's eyes as having 'an intensity stronger than rage and brighter than joy'? Who but Elizabeth Marshall Thomas would take in a wild mouse, 'sort of like helping a relative'? This is a book to cherish, full of enlightenment, curiosity, and admiration for all things animal. I loved it."

—PATRICIA MCCONNELL, author of *The Education of Will*

Tamed &
Untamed

*Close Encounters of the
Animal Kind*

Sy Montgomery
Elizabeth Marshall Thomas

Foreword by Vicki Constantine Croke

Chelsea Green Publishing
White River Junction, Vermont

Chelsea Green Publishing is committed to preserving
ancient forests and natural resources. We elected to
print this title on 100-percent postconsumer recycled
paper, processed chlorine-free. As a result, for this
printing, we have saved:

80 Trees (40' tall and 6-8" diameter)
36 Million BTUs of Total Energy
6,927 Pounds of Greenhouse Gases
37,566 Gallons of Wastewater
2,515 Pounds of Solid Waste

Chelsea Green Publishing made this paper choice
because we and our printer, Thomson-Shore,
Inc., are members of the Green Press Initiative,
a nonprofit program dedicated to supporting
authors, publishers, and suppliers in their efforts
to reduce their use of fiber obtained from
endangered forests. For more information, visit:
www.greenpressinitiative.org.

Environmental impact estimates were made using the Environmental Defense Paper
Calculator. For more information visit: www.papercalculator.org.

Editor: Joni Praded
Project Manager: Angela Boyle
Copy Editor: Paula Brisco
Proofreader: Laura Jorstad
Indexer: Shana Milkie
Designer: Melissa Jacobson

Printed in the United States of America.
First printing September 2017.
10 9 8 7 6 5 4 3 2 1   17 18 19 20 21

Our Commitment to Green Publishing

Chelsea Green sees publishing as a tool for cultural change and ecological stewardship. We strive to align our book manufacturing practices with our editorial mission and to reduce the impact of our business enterprise in the environment. We print our books and catalogs on chlorine-free recycled paper, using vegetable-based inks whenever possible. This book may cost slightly more because it was printed on paper that contains recycled fiber, and we hope you'll agree that it's worth it. Chelsea Green is a member of the Green Press Initiative (www.greenpressinitiative.org), a nonprofit coalition of publishers, manufacturers, and authors working to protect the world's endangered forests and conserve natural resources. *Tamed and Untamed* was printed on paper supplied by Thomson-Shore that contains 100% postconsumer recycled fiber.

Library of Congress Cataloging-in-Publication Data
Names: Montgomery, Sy, author. | Thomas, Elizabeth Marshall, 1931– author.
Title: Tamed and untamed : close encounters of the animal kind / Sy Montgomery and Elizabeth
    Marshall Thomas ; foreword by Vicki Constantine Croke.
Description: White River Junction, Vermont : Chelsea Green Publishing, [2017] | Includes index.
Identifiers: LCCN 2017023243| ISBN 9781603587556 (paperback) | ISBN 9781603587563 (ebook)
    | ISBN 9781603587730 (audiobook)
Subjects: LCSH: Animals. | Animal behavior. | BISAC: NATURE / Animals / General. |
    NATURE / Essays. | PETS / Essays. | NATURE / Animals / Wildlife.
Classification: LCC QL45.2 .M66 2017 | DDC 591—dc23
LC record available at https://lccn.loc.gov/2017023243

Chelsea Green Publishing
85 North Main Street, Suite 120
White River Junction, VT 05001
(802) 295-6300
www.chelseagreen.com

MIX
Paper from
responsible sources
FSC® C013483

In deep, loving memory of
Molly, Tess, Sally, Pearl, Sundog, Georgia, Shelah,
Misty, Maria, Cokie, Suessi, Fatima,
Windigo, Viva, and Miska.

Contents

Dogs and Cats

Wild Animals

Foreword

---◆---

O utside of gothic works of fiction set in Transylvania, we
rarely read of enduring friendships that have been initi-
ated by a bite. But that is exactly how nature writers Sy
Montgomery and Elizabeth Marshall Thomas—the two extraordinary, quirky, and iconoclastic women whose essays are collected
here—formed their attachment to one another.

Liz and Sy met more than thirty years ago, within months if
not weeks of Sy moving to New Hampshire, just minutes away
from Liz. Sy was a journalist, writing often about wildlife and soon
to embark on her first book, on great apes and the women who
studied them. Liz had written classic accounts of life among the
San (or Bushmen) hunter-gatherers in the Kalahari Desert as well
as novels set in Paleolithic times. As a keen observer of animals,
she had also been helping researcher Katy Payne study elephant
bioacoustics. So when Sy's husband, author Howard Mansfield,
saw an article about Liz in a local newspaper, he urged Sy to get in
touch and, before long, Sy was interviewing Liz about the emerg-
ing knowledge of how elephants communicate.

As soon as Sy and Liz sat down together, the two women, who still live in neighboring towns, found common ground talking about the natural world. The discussion that day might have begun with elephants, but it inevitably moved on to lots of other species, including ferrets—Sy's pet ferrets to be exact. Liz wanted to meet them in person. Sy, eager to oblige, escorted Liz to her house, though she worried that the timing wasn't optimal. Sy was just back from six months in Australia, and the pet sitter had turned out to be allergic to ferrets. That meant they weren't very used to being handled. Sure enough, one of the animals did, in fact, sink his pointy teeth into Liz. Sy apologized and started to explain, but she found her defense was completely unnecessary. Liz said—emphatically—that she didn't mind at all. "She *really* didn't mind being bitten by a weasel," Sy says. And that's when it hit her. "I knew we were soul mates."

Since that day in New Hampshire, Liz and Sy have traveled together to Costa Rica, where they mist netted bats with Bat Conservation International; to Maine, where they followed moose cows and calves on foot in the woods; and most recently to Tanzania, to track the wildebeest migration with another illustrious "neighbor" of theirs, ungulate expert Richard Estes. Liz and Sy are in each other's lives and in each other's thoughts. Ipso facto, the prolific writers are in each other's books. Sy has "a cameo appearance" in Liz's game-changing bestseller *The Hidden Life of Dogs*—as the two happily forgo souvenir hunting in the capital of Costa Rica in order to track a female dog in heat. Liz is a hero in Sy's *The Good Good Pig*, when she figures out how to best comfort Sy's distraught and ailing dog Tess during one of Sy's absences, bringing an old barn coat belonging to Sy for the dog to cuddle up with. "I think I've quoted or included a scene with Liz in almost every book I've written for adults since we met, and she's mentioned me in every nonfiction book she's written since *The Harmless People* and *Warrior Herdsmen*. We read and critique each other's drafts, and we talk almost every day."

The essays here are mostly collected and adapted from their joint column in the *Boston Globe*, the newspaper that brought me together with these remarkable women. It's where I worked as an editor and writer starting in the late 1980s, when Sy was a regular contributor to our Science section. And it was for my own column there, "Animal Beat," that I interviewed Liz several times. When I discovered somewhere along the way that they were friends, I wanted in. Lucky for me, inclusion comes naturally to them. In fact, it is a hallmark of their work and lives. They've always rejected any kind of reflexive exclusion, scientific or personal. Sweet as they are, there's a real saltiness to their skepticism. They are, one might say, the kettle corn of nature writers.

And their concept of inclusion is pretty darn inclusive—stretching across life-forms. Both Liz and Sy are iconoclasts writing with rare insight and nuance about the many ways we humans have attempted to separate ourselves from nonhuman animals—in our language, in our taxonomies, and in our expectations. The two have rigorously poked holes in some widely held but flawed assumptions about human superiority, and they often do it with humor. When I first interviewed Liz in 1993, we talked about the stealthy power of language to reinforce old notions of "other." Why is it that we humans experience "love" but allow nonhuman animals only "pair bonds"? How can a dog giving birth to puppies be an "it," but a woman with a baby is a "she"? Some readers had objected to the fact that Liz referred to the relationship between two of her huskies, who were devoted to one another, as a marriage.

Liz was a pioneer. "Excluding miracles," she told me at the time, consciousness is something "we acquired through our long mammalian past."

"Are we the only creatures to think and love and feel?" I asked her.

She scoffed. As she had written in *The Hidden Life of Dogs*, then just released, "thoughts and emotions have evolutionary value." In other words, being able to learn means being able to deal with life's variables more effectively. Then as now, Thomas insists that it is sheer arrogance for humans to assume we are the only mammals to have developed consciousness; the only mammals to love; the only mammals to feel empathy; the only mammals to possess morals.

Was she out of bounds with these thoughts, I prodded. She had one reply to those who would say so: "Fiddle-dee-dee."

In this same talk with Liz, I asked her about the great ripples of controversy her rather slim book had set in motion. "Well," she said, flashing that huge smile I love so dearly, "I didn't write the book to be artistic, I wrote it to be Messianic!"

Sy shares this worldview that questions a human-centric sense of superiority. She's expressed it often in writing and in person. In a conversation I once recorded with her, I asked her about it. She, too, spoke of evolution and the fact that we humans owe our emotions and morality to our fellow animals. Limiting our friendships only to other humans is as absurd as insisting on a poor diet made up of only one item, she declared: "The idea that we're only supposed to have friendships with one species among the thousands and thousands . . . that's crazy. If somebody said, 'I only eat one food. I eat nothing but nachos. I never eat anything else.' It'd be crazy!"

Well, those nachos bring us back to that long-ago bite of the weasel. Liz wasn't being generous when she shrugged off the incident. It's just that she understood. She saw the incident from the perspective of that other being. And the understanding Sy and Liz bring to their love of fellow creatures includes a deep and kind understanding of fellow humans. I can testify. And it is that unique understanding that shows up in the essays collected in *Tamed and Untamed*.

In their writing and in their lives and in their remarkable friendship, Liz and Sy break down false barriers and carry us closer to our fellow creatures. They help us see the connections. It reminds me of something Helen Macdonald said in her beautiful book *H Is for Hawk*: "Wild things are made from human histories." If that's the case, how lucky we are to read the work collected here, to let the braided essays of Sy Montgomery and Elizabeth Marshall Thomas help move us closer to being better humans—by making wild things what they should be: *our* soul mates.

—*Vicki Constantine Croke*

PART ONE

· · · · · · · · · · ◆ · · · · · · · · · ·

Animals and People

*A*t the delightfully retro Peterborough Diner, where locals in our corner of New Hampshire often gather for lunch, I always order the same sandwich: the "Meatless Grinder." I always thought it hilarious that the restaurant defined this menu item—cheese, tomatoes, and lettuce on a hoagie roll—by what it was not. (They don't offer "Sauerkraut-Free Jell-O" or "Sugarless Sausage.")

Concentrating on the meat-free aspect of cheese seems as silly as referring to all animal species but one as "nonhuman animals." Humans always seem to get separated out.

Why do we do this? Humans are the only ones who will read these sentences. We're also the only ones who wear hats. But the list of attributes once thought to be unique to our species—from using tools to waging war—is not only rapidly shrinking but starting to sound less and less impressive when we compare them with other animals' powers. Spiders grow new limbs. Octopuses change color and shape. Insects and amphibians metamorphose from one distinct form to another. Human accomplishments pale!

One reason we created this collection of essays was to put humans back into the animal world and bring animals into the human world—where we all belong.

Think of it: For all but the last few moments, evolutionarily speaking, of our existence as a species, humans have been hunter-gatherers. We depended directly upon our observations of the natural world—the real world—for everything: food, shelter, clothing, medicine, even art, worship, and inspiration. The natural world is where our kind perfected "the wholeness of all we think of as culture," wrote Paul Shepard, the scholar of human ecology. And humans, as we now know, are not the only animals with culture by a long shot, as you'll see when you read the stories that follow.

How different are we from other creatures? Humans are so closely related to apes you can share a blood transfusion from a chimp. We share 90 percent of our genetic material with all placental mammals (and 40 percent with a banana!). Even the word *person* does not derive from the single meaning "human." *Person* comes from the word for *mask*, as in the Christian mystery of "God in Persons Three" (Father, Son, and Holy Spirit). A *person* means merely one of the many Masks that God wears in this world—animal or human. This truth has long been recognized in many cultures, particularly indigenous societies. Many of these tribes tell creation stories that portray animals as the First People.

In mythologies throughout the world, the theme reemerges: Animals nurture and inspire us. From Russia, Turkey, Liberia, India, Chile, and Greece, we find stories of animals who adopt human babies and raise them in their world. We read of monkey boys, gazelle girls, even an ostrich boy. From the Roman Romulus and Remus, the human twins raised by wolves, to the Sundarbans' Bonobibi, the orphaned-girl-become-goddess rescued by wild deer, our kind honors a kinship between humans and animals—and the special powers accorded to humans raised by our wild kin.

Our fellow animals also sometimes frighten and repel us. But even this can be instructive and often tells us more about ourselves than the objects of our fear. The following essays explore some of the ways we interact with fellow species, often in surprising ways. But all of them affirm the fact, told to us by both evolution and our sacred creation stories, that we belong together with our fellow animals, and without them, we cannot be whole.

—*Sy*

Animals as Teachers
and Healers

— Sy —

*M*y father was my hero. He was an army general who had survived the Bataan Death March. As a child I got in trouble in Sunday school for saying I loved him as much as Jesus. When my dad was dying of cancer, my husband knew of only one thing that might help ease my sorrow: a sickly, runty piglet.

All my love couldn't cure my father's illness. But I could love this little pig back to life. All he really needed was a little TLC, some wormer, and some slops. Well, a lot of slops. But our little pig—who grew to be a very big pig—gave as good as he got. Christopher Hogwood (we named him after the famous conductor, as we both were lovers of early music) became not only a beloved member of my family but my teacher and healer as well.

Animals have been recognized as mentors for millennia. "The animals are great shamans and great teachers," the mythologist Joseph Campbell insisted. Any animal, he said, "may be a messenger . . .

or one's personal guardian come to bestow its warning or protection." Among the North American Oglala, a person on a vision quest seeks an animal teacher. It might be a bear, since bears know the healing powers of plants. (Scientists have now proved that bears indeed do use plants as medicines, including willow bark as aspirin.) Or it might be an eagle, since eagles can see all that happens (and in fact, an eagle can see an animal as small as a rabbit more than two miles away).

These days, modern readers are catching up with what native people always knew. So many recent books have been written about animals as healers that it's almost a literary genre. While the motif is familiar, these stories retain the power to surprise—because the ways an animal can rescue, instruct, comfort, and empower us are virtually endless.

In Lissa Warren's *The Good Luck Cat*, a cat leads the author out of the abyss of grief. The book's heroine, a mischievous seven-pound blue Siamese named Ting, was really Warren's dad's pet, purchased as a retirement companion. When her beloved dad dies suddenly of a heart attack, for Warren and her mom the cat he adored becomes a feline embodiment of the man. But then Ting herself develops serious heart trouble. And the book's subtitle, *How a Cat Saved a Family, and a Family Saved a Cat*, tells only part of the story, because there's yet another twist: After Warren had begun writing this book, her left leg went numb, and then her left arm, then her face—and chapters she had never expected to write detail how Ting's healing powers get called into play again.

In a very different memoir, *H Is for Hawk*, when Helen Macdonald's dad dies, she turns not to a soft, purring cat but to "the bastard offspring of a gleaming torch and an assault rifle." That's her description of a goshawk, the most voracious and unpredictable bird known to falconry. The young female she procures, Mabel, is everything a grieving human is not: Boiling with life, the red-eyed raptor "was a fire that burned my hurts away." But eventually the fiery goshawk leads the author to what she considers a kind of madness, where she retreats from the human world and enters another, older and wilder, a world where she realizes she does not belong.

Elisabeth Tova Bailey, too, leaves the human world, but it's a strange, debilitating illness that transports her there. Beset in her active, busy thirties with a rare pathogen, she can no longer stand upright. Bedridden and sapped of energy, "each moment felt like an endless hour," she writes. She receives as a gift a pot of wild violets with a woodland snail living in it.

In her earlier, healthy life, she wouldn't have given a snail any thought. But in her lovely book, *The Sound of a Wild Snail Eating*, she details how her illness gave her the patience to observe and appreciate the snail's mysteries and charms. She makes some genuine scientific discoveries as well as a number of personal ones—including that "my snail was just as aware of the details of its world as I was of mine."

Then something happens to break the spell. Bailey gets better and discovers she no longer has the patience to observe her snail. She returns to the large, boisterous, human world. But she does so enriched by her glimpse of a slower, softer world that exists beside our own.

For the Oglala the spirit of the animal teacher actually enters the body of a person on the vision quest, so that the animal forever after becomes part of his strength. The Oglala got it right: Animals become part of us, restore us, and remake us—and they give us the power to restore and remake others when we share their stories.

Your Brain on Pets

— Sy —

I once faced a sickening defeat. The only thing keeping me from swimming with octopuses in the wild was a scuba certificate. The only thing keeping me from seeing octopuses in the wild was a scuba certificate. After a day and a half of an intensive scuba class, diving too deep, too fast produced pressure in my ears, causing dizziness and nausea. I was forced to quit. Next I realized I was too vertiginous to drive home.

Despairing, I lay down on the blanket that protects our car's upholstery from our border collie's dirty paws. As I inhaled Sally's scent, calm washed over me. Within a half hour, the dizziness eased enough for me to drive.

We animal lovers have long known that no matter what life may bring—sickness, sadness, or radiant health—pets make us feel better. Numerous studies have documented astonishingly wide-ranging effects. Cat owners enjoy a 30 percent reduction in heart attack risk. Watching swimming fish lowers blood pressure. Stroking a dog boosts the immune system. Now researchers can explain the source of our companion animals' healing powers: Our pets profoundly change the biochemistry of our brains.

"This is science that supports a truth the heart has always known," Meg Daley Olmert writes in her book, *Made for Each Other,* a synthesis of more than twenty years of work on the biology of the human-animal bond. She singles out one neuropeptide: oxytocin, a brain chemical long known to promote maternal care in mammals.

Oxytocin levels rise in a mother's brain as she goes into labor and produce the contractions that deliver the baby. Once her infant is born, just the sight, smell, or thought of the baby is enough to trigger milk letdown (a fact that has caused many a new mother to ruin a blouse). Humans have known for millennia that this affects animal mothers, too: Ancient Egyptian tomb art shows a kneeling man milking a cow with her calf tethered to her front leg.

But oxytocin's powers are not, as once thought, limited to mothering or triggered only by labor. Nor is it confined to females, to mammals, or even to vertebrates. Even octopuses—who not only lack breasts but die when their eggs hatch—have a form of oxytocin, called cephalotocin.

Oxytocin causes a cascade of physiological changes. It can slow heart rate and breathing, quiet blood pressure, and inhibit the production of stress hormones, creating a profound sense of calm, comfort, and focus. And these conditions are critical to forming close social relationships—whether with an infant, a mate, or unrelated individuals—including, importantly, individuals belonging to different species.

In a study published in the *Proceedings of the National Academy of Sciences,* Japanese researchers sprayed either oxytocin or saline solution into the nostrils of dogs, who then reunited with their owners. The owners were told not to interact with their dogs, but the pets who inhaled oxytocin found their people impossible to ignore. Statistical analysis showed the oxytocin inhalers were far more likely to stare, sniff, lick, and paw at their people than those who had saline solution.

Oxytocin is not the only neurotransmitter companion animals call forth from our brains. South African researchers showed that when men and women stroked and spoke with their dogs, the people's blood levels of oxytocin doubled. But the interaction also boosted levels of beta-endorphins—natural painkillers associated with the "runner's high"—and dopamine, known widely as the "reward" hormone. These neurochemicals, too, are essential to our sense of well-being. A later and larger study by University of Missouri scientists also documented that petting dogs caused a spike in the people's serotonin, the neurotransmitter that most antidepressants attempt to elevate.

So it's no wonder that pet-assisted therapies help troubled children, people with autism, and those suffering from drug addiction and posttraumatic stress disorder. Pets help normalize brain chemistry.

"By showing how interacting with pets actually works," says the Missouri study's lead author, Rebecca Johnson, "we can help animal-assisted therapy become a medically accepted intervention"—one that could be prescribed like medicine and reimbursed by insurance.

All animals appear to have cells directly under the skin that activate oxytocin in the brain. So gentle touch—from grooming your horse's coat to making love with your spouse—is a powerful trigger. But so is simply thinking about someone you love, whether it's a person or a pet. And in fact, a small study at Massachusetts General Hospital published in October 2014 found that MRI scans of women's brains lit up in the same areas when shown pictures of their pets as when shown pictures of their children.

But here's the best part: It's mutual. We effect the same physiological changes in our pets as they do in us. As I lay on that blanket in our car, soothed by Sally's scent, I remembered how my best human friend, Liz Thomas, once quelled desperation and fear in another border collie named Tess, Sally's beloved predecessor. I was away tending to my dying mother when Tess, a rescue with separation anxiety, suffered a strokelike illness. For the first time in her life, she was confined overnight at the vet's. Liz knew just how to help. She came to our house, retrieved my barn coat, and took it to Tess's hospital cage. Tess inhaled my scent, and instantly her ears folded and the terror fell from her face. She let out a sigh and relaxed.

Is ESP Possible?

— *Liz* —

*I*s there such a thing as ESP, or extrasensory perception? Most scientists doubt it, and rightly so. Experiments were tried, but none could demonstrate it. One experiment, for instance, involved two people in isolation booths, one turning up cards from a pack while the other tried to identify them using ESP. The results were disappointing.

Even so, I've had four ESP experiences, all generated by powerful emotions that the sight of a playing card would not arouse. Three times a dog was involved and perhaps was the vector. So humor me, please, while I tell my stories.

One dark night I was accidentally locked in a museum where I'd been working. I didn't like the museum: I was young at the time and living with my parents, and was afraid of some mummified corpses that I knew were stored in its basement. Suddenly the lights went out. Terrified, I groped my way in pitch darkness to the door, but it was locked. I heard footsteps. The mummies were walking! In panic I tried other doors until I found one that opened, and I ran out into the night. We lived nearby, and my mother was in front of the house, badly frightened. She had suddenly felt that something terrible was happening to me and had come outside to look. I wasn't endangered—the footsteps were the janitor's as he closed the building. But never before had I experienced such fear, and somehow my mother caught it.

Panic over another event was transmitted either by our dog Ruby when our dog Sheilah was killed or by Sheilah herself as she suffered. It happened in our driveway, evidently when two men in a truck unknowingly ran over Sheilah with their trailer-rake, which dragged her. No one saw this but Ruby.

As for me, I was fifteen miles away, driving to a fabric store, thinking about a quilt I was making, when suddenly I felt a sense of terror, of emergency, of terrible trouble at home. I tried to calm myself and think about the fabric, but urgency overwhelmed me. I made a U-turn, sped back to our house, and saw Sheilah's body on the lawn. She'd been found dead on the road. I later investigated the drag track with another dog, Pearl, who explored the track with her nostrils, her hair bristling with each terrible discovery. That's how we knew what happened.

In another experience, my husband and I were living in an apartment with two dogs. At the time, the younger dog, Violet, wasn't completely housebroken, and because we were planning to go out, I considered leaving her on the balcony. I imagined myself taking her to the balcony and shutting her out there, but I didn't physically move because I wasn't actually going to do this—it would have been extremely unkind. It was simply a thought that had crossed my mind.

My husband was in the same room, taking a nap. He half-opened his eyes. "Don't leave her out there," he said.

He thought he'd heard us on the balcony, heard the door shut, and heard me come back inside alone. Did he get this from me or from Violet? It might have been from Violet. Some scientists believe that animals think in visual images. I normally think in words, but this time I had pictured what I thought of doing. I felt no emotion at the time, but Violet would have, if she caught it.

Must an emotion be powerful to transmit ESP? One lovely summer day I waited by our car with our dog, Sundog, while my husband went into a store. Sundog and I were sitting on the hood of the car enjoying the sunshine when suddenly, for no apparent reason, a crushing, desperate sadness overcame me. Sundog was beside me, suddenly alert, sitting up straight, looking at the store, his body tense, his ears raised, his eyes wide—as if he thought that something had happened in the store.

After a while my husband came out, his head low, walking slowly. He'd learned that the storekeeper had driven the ambulance that came for our teenage daughter who, many years earlier, had been hurt in a terrible accident. My husband hadn't known who drove the ambulance, and the memory of the event overwhelmed him. It seemed that Sundog picked that up.

Our daughter survived the accident although she was disabled, and by then was an adult. We lived in New Hampshire, she lived in Texas; Sundog hardly knew her; and the accident had happened before he was born. All he could feel, if indeed he felt something, was my husband's grief. Like a lightning bolt it transmitted to me, but only the terrible grief. While feeling it I made no association with our daughter or the accident.

Coincidence might explain these events, and ESP may never be proved. All that can really be said, at least for now, is that if we have it, other animals do, too.

Thunder

— Liz —

Summer storms are upon us, and with them, thunder. Most of us who live with dogs have noticed that they don't like thunder. One of mine was a capable sled dog who before she came to me had lived at the end of a chain in a village in northern Canada where she learned all about terror, mostly from her former owner who beat his chained-up dogs when he was drunk. Yet for all the terrible things she had seen, she was fearless—except when it came to thunder. She'd hide in the bathroom, squeezing herself behind the toilet, trembling so hard her teeth chattered.

Our neighbors' dog was once home alone when thunder started, and when her owners came back they couldn't find her. The storm was still raging while they searched the house, calling her, until they noticed a paw sticking out of a pile of laundry in a basket. The dog had climbed inside and buried herself under the clothes. In my experience, dogs hide in closets or other small spaces but don't cover their bodies, so this surprised me, and shows, I think, the depth of that dog's concern.

Thunder

My dogs were not afraid of bears or any animal, not even of skunks or porcupines despite some sad experiences. One of our dogs noticed that something burning on the untended stove had set the kitchen curtains on fire, and she knew we were in danger. But she didn't run or hide. Instead she faced the flames and barked so loudly and insistently that at long last we came to see what was wrong. The fire alarms had not sounded—who knew how far the fire would have spread before they did? This brave dog saved our house and all who lived there. But even she was terrified of thunder.

I find it hard to reassure a dog when thunder is rolling. They let you stroke them and pat them and tell them it's okay, but they don't believe you. Are you so dense, they wonder, that you don't understand what's happening? It seemed to me that all dogs feared thunder, that it must be hardwired in them for some reason, and the fact that the noise itself does no actual harm was incomprehensible to them.

Or that's what I thought until a Chihuahua joined our family. He was two years old, weighed nine pounds, and came from a shelter. Our cats are bigger than he is. When we're outdoors he stays nearby, and when we go to bed at night he likes to be with me under the covers. All you can see is a little lump in the bed. It's him, curled up and sleeping peacefully.

One day, soon after he joined us, a thunderstorm started. It was right overhead, so I thought he'd be terrified and I'd have to console him until it was over. But instead he ran to the door and looked up at the sky, barking ferociously and showing his teeth.

The thunder got scared and stopped immediately. A few minutes later we could hear it far away. The little dog barked and snarled again, but not as forcefully. He knew the thunder was retreating.

This little dog can handle any problem, especially those from the sky. One day he took exception to a low-flying plane heading for a nearby airport. He ran outside barking and snarling and the plane flew away fast. We think it warned the other planes, because no planes now fly low. By the time they're over us, they're up very high. The little dog looks to make sure they're staying up there, then goes on about his business.

I think of his courage when we're sleeping. I feel his warm little body at the back of my knees. I hear soft little puffs of his breath. He may be small, but that doesn't worry us. We know he'll keep everyone safe.

Animal Minds

— Liz —

I write this in my office with my two small dogs nearby, curled up together on a chair where both dream. Sometimes one will cry softly, sometimes one will jerk his legs as if he were running. Observers are prone to say, "He's chasing rabbits," even if he's never seen a rabbit, and assume the dream can be of no consequence because the dreamer is just a dog. We don't really know what dreaming does for us, but whatever it is, it does the same thing for dreaming dogs and other dreaming animals, including birds and fish.

Virtually any mental manifestation one can think of—emotion, reasoning, learning, fact-finding, decision making, sympathy, empathy, recognition of "other," and many more—is present in all kinds of animals, certainly the vertebrates. Although some of us acknowledge a few of these mental features in our pets, too many of us still cling to outdated scientific theory and deny the existence of animal cognition in almost any form.

This wasn't always so. During our first 200,000 years as hunter-gatherers, we had to recognize the mental abilities of other species just as they needed to recognize ours, especially if they hunted us or we hunted them. But it seems that the more "formal education" we acquired, the less we understood the truth. PhD philosophers spent lifetimes trying to define the fundamental difference between the minds of animals and the minds of people when all along there is no fundamental difference and never was. And until very recently, given that no scientific proof existed either way, most of the scientific community chose to assume that animals did not have consciousness or emotions or thoughts.

A tsunami of evidence now refutes this. Fascinating films, books, and papers are shedding new light on the conscious lives and mental abilities of animals. Even a paramecium proved that it could learn. What is a paramecium? It's a tiny, oblong single cell that swims. You can't see it without a microscope. Yet the paramecium in question learned to avoid a certain kind of light. True, there's still some distance between that paramecium and Stephen Hawking or Albert Einstein, but the paramecium was on the same track.

A breakthrough in this area was recently published—a book by Nicholas Dodman called *Pets on the Couch*. Even the cover promises realism—it shows a little dog sitting on the seat of a couch, a macaw perched on the back of the couch, and the tail of a cat hiding under the couch. Dodman is a veterinarian who treats what we call "behavioral" problems but which in reality are psychological problems and, as Dodman has determined, are much the same or exactly the same as ours. Dodman has found that dogs and cats with such problems respond to pharmaceuticals prescribed for humans with similar problems. He has found this true of multiple psychological disorders, from obsessive-compulsive behavior and Tourette's syndrome (a horse had Tourette's syndrome) to Alzheimer's disease, depression, and posttraumatic stress. This wide array of problems stems from different kinds of mental disorders, and in my opinion, the fact that the same pill fixes the same symptom, whether in a person or a horse, shows us the most important thing we need to know about animals: They are more like us than we thought. Their organs such as their hearts, lungs, and kidneys are pretty much the same as ours in function and appearance—we knew that. What we didn't acknowledge, but what is not surprising, is that their brains are, too.

Not everyone sees this yet. Dodman says: "Despite Darwin, despite Goodall, despite Temple Grandin, and so many others, we still find ourselves having to apologize in scientific circles for ascribing the power of thought to animals." This should end. It's time to file that ill-considered, antiquated "scientific" theory with the flat-earth theory and acknowledge what's been under our noses for hundreds of thousands of years.

As for me, I'm with Dodman. I know, for instance, that birds, mammals, fish, certain mollusks, and even insects think and feel in much the same way we do. Even so, we kill them and eat them, and although I used to eat whatever was served, and I cooked "regular" meals for my family and guests, I no longer do so. Like Sy, who has long been a vegetarian, the last thing I want to do is eat an animal. I look at that piece of meat lying on my plate and wonder who he or she was. Who were her parents? Her siblings? Where did she spend her childhood? Was it pleasant or stressful? What did she like to do or think about, and what things did she remember? What form of captivity did she suffer and what were her last hours like? And I'm supposed to cut off a piece of her corpse and put it in my mouth?

Octopus Love

— Sy —

*T*he lights were low. The roses were tied with satin ribbon. Barry White's sexy bass throbbed on the sound system: "I can't get enough of your love, babe." It was Valentine's Day, and I had big plans to celebrate: I had flown to Seattle to watch two giant Pacific octopuses mate.

Every February 14 for more than a decade, the Seattle Aquarium has hosted the "Octopus Blind Date." It's surprisingly popular with children. The year I went, 150 sixth graders, 88 second graders, and kids as young as five from at least five other elementary schools lined up in front of the three-thousand-gallon, two-part tank, strung with heart-shaped red lights, waiting for the moment that Rain, the sixty-five-pound male, and Squirt, the forty-five-pound female, would meet. Everyone was eager to see what they would do.

It's a fraught moment. Despite the plastic roses floating in the tank and the romantic music over the PA, not every blind date works out. Most of their lives, giant Pacific octopuses are solitary. One year one octopus ate the other. (Happily this didn't happen in front of the public, but after the visitors had gone home.) Another year the female was scared of the male; at his approach she inked and fled.

The aquarium's lead invertebrate biologist, Kathryn Kegel, estimated there was a fifty–fifty chance that Rain and Squirt would hit it off. If there was a problem, she and another diver, clad in dry suits, would try to separate them. But, she admitted, "There's too many arms to do much about it, though."

With those sixteen arms, plus 32,000 suckers and six hearts (each octopus has three) beating as one, octopus sex would seem to offer possibilities that leave the *Kama Sutra* in the dust. Not so—at least compared with other seagoing invertebrates. Take the sea slug, *Chromodoris reticulata*, which lives in shallow reefs around Japan. All have both male and female sex organs—and can use them both at the same time. By contrast, most octopuses usually mate in one of two familiar ways: male on top or side by side.

"Our divers are going to encourage Squirt to come out and meet Mr. Rain," the aquarium emcee announced to the crowd. Kegel and the other diver then lifted up a Plexiglas barrier that had been separating the two halves of the tank. But Squirt didn't need much encouragement. Purposefully she flowed from her side of the tank, crawling along the bottom toward Rain, who was sitting on a rock wall at the opposite end.

With excellent eyesight as well as sensitive chemoreceptors all over his body, Rain knew very well that she was coming. As she approached, Rain changed color from gray to red—the color of excitement. Squirt stretched two arms toward him, and at her touch, Rain poured down the side of the rock wall. He raced into her arms. She flipped upside down. The two embraced, mouth-to-mouth, thousands of suckers mutually touching and tasting one another. Both flushed red with emotion. And then they were still.

Shortly thereafter the children decamped for their buses. Many of the kids appeared baffled. If human sex was incomprehensible, octopus sex was unfathomable. Aristotle explained octopus mating this way: "The male has a sort of penis on one of his tentacles . . .

which it admits into the nostril of the female." That's essentially correct: He uses a specialized arm to place a single, foot-long spermatophore into the large opening on the side of what looks like her head. But what most people think is the octopus's head is really the mantle, containing most of the organs.

After the kids left, the two octopuses stayed at the bottom of the tank, not moving. Rain's body covered Squirt's completely. Rain's color turned paler and paler. Finally he turned completely white—the color of a relaxed octopus.

The octopuses were not moving, so I watched and listened to the people. Two guys with arms around each other gazed into the tank, watching solemnly and in silence. An elderly couple walked by, the wife leading the husband, who used a walker. "They're mating, Leo!" she told him. "It's a very beautiful experience!" The murmurs from the humans quietly watching these marine invertebrates—creatures who last shared an ancestor with us half a billion years ago—were tinged with tenderness:

"They're so peaceful."

"He looks happy."

"They're beautiful. Just gorgeous."

"So dear. The dear, sweet things."

It would be difficult to imagine a creature more different from a person than an octopus: We are creatures of the land, they of the sea. We are full of bones, and they haven't any. They can taste with their skin and squeeze their baggy bodies through tiny openings. We mate early in life and may give birth year after year. Octopuses mate at the end of their lives, and the female lays eggs—up to 100,000 of them—all at once.

And yet, on that Valentine's Day, the octopuses and people seemed to share the sweetness of the occasion: a celebration of the pleasures of love.

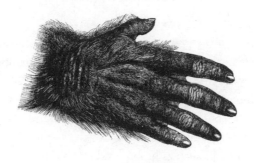

Bushmeat

— *Liz* —

As we all know, due to our damaging environmental practices hundreds of species are facing extinction. Among other factors, we blame climate change and habitat destruction, but sadly one major threat to wildlife is almost completely ignored. A few valiant scientists and science writers have tried to point this out, most recently author Dale Peterson in *Where Have All the Animals Gone? My Travels with Karl Ammann.*

The answer to that question, writes Peterson, is that the animals have often gone to fancy hotels and restaurants, mostly in Africa and Asia, and from there to people's dinner plates. Together with Karl Ammann, a Swiss wildlife photographer, Peterson traveled in these continents to observe and document the destruction. A book they prepared jointly, *Eating Apes*, describes their findings, but in both books the facts were so appalling that publishers rejected their work, believing their readers did not want to see a dismembered chimpanzee, or an elephant's trunk cut into circular slices as prepared for the table, or a bowl of wild animal soup with the animal's little paws floating in it. Ammann's photos were not taken in the homes of the poor, where food would be scarce. They were taken where bushmeat is an expensive and fashionable delicacy.

Of course, African and Asian cultures are not alone in serving eye-catching meals—witness our roasted pig cadaver with an apple in her mouth—but the issue doesn't revolve around cuisine. Instead, like many other problems, the issue revolves around money.

Large animals such as hippos and elephants are favored foods, and great apes are highly prized. People prefer to dine upon the hands and feet of great apes (boiled), but they also eat the other parts with relish and are willing to pay for it, so the corpse of one adult gorilla meant for the table can bring about $38,000, while the corpse of a cow might bring about $1,600 and the corpse of a pig about $700. This would be one reason why restaurants and hotels are promoting bushmeat, and surely the poachers are glad to provide them. Not only that, but other forms of meat aren't as readily available in Africa and Asia as they are in other parts of the world. For instance, the cattle who provide beef require long-term care and a grazing area, and to Africa's pastoral people are more valuable than money. These people don't eat or sell their cattle: The cattle themselves are wealth. In contrast, wild animals take care of themselves, and to procure one for bushmeat requires no more than a bullet.

Interestingly, in Europe and the United States, criticism of bushmeat is sometimes regarded as cultural disrespect, as if everyone who eats bushmeat lived off the land or traditionally hunted gorillas and elephants for the table. In much of Africa, most farmers or pastoralists don't often hunt for meat, and those who once did were hunter-gathers who for the most part hunted large antelopes, none of which have gone extinct due to hunter-gatherer activity. Eating bushmeat on a grand scale has only recently become fashionable, so we're speaking of a new development that is causing the mass extinction of elephants, hippos, bonobos, chimps, and gorillas, to name but a few of the victimized species.

In their travels, Peterson and Ammann found dozens of species on the menu, from rats to pythons, elephants, and hippos, but perhaps most dangerously, bonobos, gorillas, and chimpanzees. Like us, their relatives, these great apes reproduce slowly, with long gestation periods, usually for just one offspring at a time. The bonobo population is now reduced so seriously that it will almost certainly vanish forever.

Can anything be done? Opponents of bushmeat could point out that eating chimps, gorillas, and bonobos is a form of cannibalism, but probably that would deter only a few. Many nongovernmental organizations now actively discourage the use of bushmeat and are educating people not only about endangered species but also about diseases one can get from eating bushmeat. The United Nations recommends insects as a source of protein, and it seems that some insects are quite palatable. Years ago I ate an African ant whose species was named for honey (albeit in the Ju/'hoan language), so I know this is true. Insects reproduce prolifically, not expecting all their hatchlings to grow to adulthood, and their appearance on menus wouldn't dent their populations. It's said that sago grubs taste like bacon and that cockroaches are less fattening than beef. Eating them won't save the world, but it could help.

Cecil the Lion

— *Liz* —

Most of us saw the reports of the lion named Cecil who was illegally shot with a crossbow in Zimbabwe by an American dentist named Walter Palmer. This caused such international fury that the president of Zimbabwe called Palmer a poacher, and President Obama received more than 160,000 signatures petitioning that he be extradited to Zimbabwe for trial.

In contrast, a television pundit told viewers that more than a hundred lions had been killed that same year in Zimbabwe and this was okay because big-game hunting produces a meaningful amount of the national income. According to the pundit, the reason Americans were so excited about this particular lion was because he had a name.

I don't think so. As for the national income, the pundit was wrong. Hunting contributes little to the national income—not nearly as much as tourists do when on safari to look at wild animals. Palmer allegedly paid $50,000 for the privilege of killing a lion, a normal price for big-game hunters. Such money goes first to the hunting-safari companies, most of which are owned by white expatriates who organize the hunts and serve as guides. The hunting-safari companies then reward the appropriate government officials for ignoring illegal practices, as was disclosed by the ecologist Craig Packer, who studied not only lions but also the corruption of government officials in Zimbabwe and learned what happened to important amounts of the white hunter's money. The corrupt officials don't share their profits, and the national income doesn't rise.

As for the outraged American public, we weren't told about one hundred lions, only about one. We were shown a video of a huge dying animal lying on his side with what looks like a long metal pipe in his chest, trying to move, trying to breathe, his eyes open and searching, surrounded by excited people happily taking photos, and we were outraged that anyone could take pleasure in this tragedy.

And by the way, the dentist gets no credit for using a bow and arrow. The pipe in the dying lion's chest was, of course, the "arrow." To hunt a lion with a bow and arrow rather than with a high-powered rifle might make Palmer seem brave, but he wasn't using an archer's bow—his crossbow was a powerful, complicated machine that fires a huge steel projectile, perfectly capable of killing a lion slowly and extremely painfully.

Shortly after the killing, comments on the lion's murder circulated on Twitter and other social media. Some people criticized those who mourned for the lion "because children are starving." Isn't it possible to feel concern for all kinds of animals, young or old, human or otherwise, especially if you don't consider animals as things? This murder of Cecil involved an intelligent, sentient being much like ourselves, one with a family and responsibilities. Those who mourned the lion understand this, and those who criticized the mourners don't.

So maybe the lion's name was important. Most animals—the vertebrates anyway—don't look like us or behave as we do or communicate in ways we understand, which is why we see them as unimportant. Perhaps a name makes them seem more important, or in other words, more like us. This gives a more accurate impression, because in every way that matters, such as having consciousness, memory, thoughts, and emotions, animals are almost exactly like us, which isn't surprising because we are animals, too.

An important book by Carl Safina, *Beyond Words: What Animals Think and Feel,* presents mind-blowing accounts of what he saw animals doing. Safina, who is highly respected for his intensive fieldwork, recorded animals displaying their consciousness, their reasoning, their knowledge, their memory, and their emotions. His accounts are exact and convincing, suggesting that all who still believe that animals lack these qualities know significantly less about other animals than other animals know about them.

Once when I was in Etosha National Park, Namibia, taking part in an elephant study, I was sitting on the ground, fixing something inside a small fenced enclosure made to protect the park wardens who patrolled on horseback. On the other side of the fence a lioness was lying on her side with her head raised, dreamily watching me work.

I yawned. She yawned, too. Amazed, I waited a moment and yawned again. She yawned again. She did this several times but stopped when she realized I was manipulating her. She had been watching me with empathy (very different from sympathy). She wanted to know what about me was like her. Since we were pretty much the same, we both found yawning catching, as I already suspected and as she discovered through casual observation. It's how animals study other animals. If our species did the same, we'd find better ways of spending $50,000.

Discarded Animals

— *Liz* —

*O*n a cold night in September 2010, a horrible person drove along the country road that leads past our house in New Hampshire, stopped the car, and threw two kittens into the bushes. Our neighbor found them—she noticed their tiny faces looking through the bushes. She took them home but couldn't keep them, so she called me and I took them. They were about three weeks old, thin, starved, and infested with fleas.

But they were lucky. Bobcats, bears, coyotes, and fishers live in our area, as do hawks and owls, and they all eat kittens when they can. The kittens were frightened but understood they'd been helped.

Of course, our neighbor fed them when she found them, but they ate again greedily when I brought them home because they had learned what starvation feels like. We feed our cats on the kitchen table so the dogs don't eat their food, and when the kittens could eat no more they explored the table briefly, then got in a small box that happened to be there, curled up together, and slept for hours in peace. They're grown-ups now, and we bless the day our neighbor found them because they so greatly enhance our lives.

We live in a small town, population about six thousand, and many of us know each other, so we soon learned that our kittens had a sister. A friend who lives about three miles away found the sister on the same day our kittens were discovered, and she can only have been discarded by the same dreadful person. The kitten was the same age and in the same condition as ours, and we feel sure they're related because all three are purebred Russian blues, suggesting that whoever threw them away is not only cruel but ignorant. Such kittens sell from $400 to $500, and anyone who pays that kind of money for a kitten won't be throwing it out of a car. Obviously, the dreadful person hadn't purchased them, but he or she could probably have sold them and bought drugs or whatever else made him or her so thoughtless. Our community is not only small but cohesive. We don't think of ourselves as cruel or stupid. When I tell others our story, they say the criminal must have come from out of state.

Perhaps, but not necessarily. One day I found a dead domestic rabbit, a Dutch rabbit, in our garage. Our dogs had killed him. I had no idea how a domestic rabbit came to be there until weeks later a woman told me that she once had a Dutch rabbit whom she didn't want so she "let him go" in our field.

This explained it. Never having spent even one day in the wild, at a loss to find himself alone and abandoned in an open field, the rabbit had come to our house in hopes of finding someone to help him. It breaks my heart to think of him trusting our species, being dumped out of a bag or a box and left alone in a strange field, not knowing where he was, seeing our house in the distance, then crossing a road and climbing a hill to reach it. It breaks my heart that I didn't know he was there. I would have helped him.

People like that woman give me the shudders. If she didn't want the rabbit but chose to release him on my land, why didn't she give him to me? She's a nice enough person in other respects, but today if I notice her at some town event, I pretend I don't see her.

Why do we have animal shelters and humane societies? Because they save animals like that rabbit and those kittens.

The people who manage the shelters don't condemn you for not wanting an animal, or because you moved to an apartment that doesn't allow animals, or because your new boyfriend or girlfriend doesn't like animals and you chose the person over the pet. I have a refrigerator magnet that shows two women talking, one saying, "He said I had to choose between him and the dog. We miss him sometimes." We instantly sense that the dog is still with her and a disagreeable gentleman is not, but too many people make the opposite decision, and the shelter is there to help whatever animal someone doesn't want. Over the past fifty years, most of my animal companions were rescues, some because they came to our house, some as strays I noticed lost and wandering and couldn't locate the owners, and some from shelters. I never enter a shelter without realizing what splendid work these institutions are doing in an almost impossible climate of need. I try to support our local shelter, I urge others to do the same, and I urge anyone with any animal to understand that it can't just walk off into unfamiliar woodlands and care for itself successfully any more than one of us could. It's like dropping the animal down a well.

A Memorial Day Tribute
to War Animals

— Sy —

My father never spoke to me about the wars. He had spent years as a POW. All he ever told me about his experience was that there were wild monkeys in the Philippines, and he had loved watching them before he was captured. As a little girl, unable to imagine war's other horrors, I hated the enemy for taking my father away from the monkeys he loved.

He was career army, and I was still in elementary school, too young to have an opinion about the conflict, when he was sent to Vietnam. But when he returned I overheard him say that in the jungles, elephants used as transport vehicles were being bombed. That made me feel sick. I think it made him feel sick, too—enough to prompt his choice to take early retirement as a brigadier instead of staying in to make lieutenant general.

That civilians die in war is tragic; that we drag other species into our conflicts deepens the sorrow. Though it's not true that animals don't, themselves, wage war: Chimps have reportedly annihilated rival chimp bands; certain species of ants raid other ant colonies and even take prisoners, whom they enslave. These are the animals' affairs. They don't get us involved. But humans, it seems, have enlisted animals in our bloodiest human-to-human conflicts since ancient times.

Roman naturalist Pliny the Elder tells us how in the first century AD, animals were used as unwitting war weapons. Live pigs were deployed almost literally as cannon fodder. They were set on fire and hurled at approaching armies in order to scatter the enemy's war elephants. The pigs, screaming in pain (at high decibel levels that can cause hearing loss), terrified even the brave elephants, who then panicked and trampled any soldier standing in their path.

Later advances in warfare exploited animals' more-than-human powers to serve us in our conflicts. At a time when most armies marched in long columns, Genghis Khan conquered an area the size of the continent of Africa with horses. Thanks to their horses, Mongol soldiers were faster and far more fearsome than mere men on foot. The sound of the animals' thundering hooves terrified the enemy, and the dust they kicked up cloaked them in frightful mystery.

By World War I both sides were equipped with large cavalry forces, with camels, mules, and donkeys serving as well, sometimes outfitted with equine gas masks. Other species were pressed into service: Glowworms helped soldiers read maps and letters in the trenches. Slugs (who, like canaries in coal mines, are exquisitely sensitive to pollutants) warned the men of mustard gas in time to put on their masks. Dogs laid communication wires. Pigeons transported messages. To this day, our navy trains dolphins to use their sonar and sea lions to employ their keen underwater senses of sight and smell to detect submarines and gather undersea intelligence. There is even an effort to implant electrodes into the brains of dogfish to control their movement and turn these sharks into underwater spies.

Animals, like human soldiers, have received honors for their bravery on the field of battle. For ferrying a message that saved hundreds of troops, a carrier pigeon named Cher Ami was awarded the Croix de Guerre, a French medal. For rescuing the wounded behind enemy lines, a bulldog named Stubby was awarded the rank of sergeant—causing the dog to outrank his owner. Elephants are credited for helping to win Allied victory in Burma in World War II. They alone were powerful and agile enough to drag huge logs from the forest and lift them into place to create instant "elephant bridges" across creeks and streams, allowing troops and tanks to cross. By war's end, elephants had built 270 of them.

Journalist and author Vicki Constantine Croke writes movingly about the wartime contributions of Burma's elephants in her national best seller, *Elephant Company*. Its hero, "elephant whisperer" J. H. Williams, spent his career, she writes, "trying to make the lives of working elephants better. Now he saw creatures who had no understanding of war being sacrificed to it." And this grieved him deeply, as he considered his elephants comrades in arms.

One of them in particular, the brave tusker Bandoola, Williams considered a brother. Williams's relationship with Bandoola illustrates a deeper reason that humans bring animals with us when we head to war. "I've always been haunted by the reports of young soldiers in Burma who called out for their mothers as they lay dying in the jungle," Croke observed when we spoke of this together. "I think the need for deep emotional connection is almost beyond reckoning, and whatever comfort these animals brought was a blessing."

No wonder that troops around the world, across the ages, have kept animals as diverse as monkeys, cats, bears, dogs, and even lions as wartime mascots—animals whose mere presence cheered the soldiers, boosting morale. "We know that it can often be easier to express affection for an animal companion than a colleague or comrade," Croke notes. "And, believe me, if you've ever been trunk-hugged by an elephant"—and she has!—"you feel the love."

By the time my father died in 1991, he had revealed no more about his wartime experience than he had during my childhood. But I hope that from his jungle prison camp, he could sometimes glimpse, if not a monkey, then a bird or squirrel—giving him some cheer, reminding him of hope and freedom.

A Failure to Communicate

— *Sy* —

A few years back a friend who works with elephants told me about an animal communicator she met, who reported she had a telepathic conversation with an aggressive zoo elephant. The communicator claimed the elephant really liked her—so much so, in fact, that she said the elephant had wanted to put his massive head in her lap.

Unbelievable! And in fact, it was.

If indeed the animal communicator had received a message from the elephant, she had grossly misinterpreted it. A child might show trust and affection by laying her head in your lap, but an elephant who does this is trying to kill you. They use their heads to squash irritating individuals like a person grinds out a cigarette butt with his shoe.

Misinterpreting an animal's motives or behavior is easy to do. We want to understand animals—and we want them to like us—and this colors our perceptions. A zoo veterinarian about to do an exam on a Tasmanian devil, a carnivorous marsupial the size of a small dog, noted with pride and relief how calm the animal seemed in her presence. "Look," she said, pointing to his wide-open jaws, "he's so relaxed he's yawning!"

It was then that the Tazzie devil bit her. What looked like a yawn was really an attempt to warn the veterinarian to stay away. The "yawn" was what's known as a gape threat, the animal's effort to advertise the power of its strong jaws and sharp teeth.

Such mistakes may lead people to conclude that an animal is mean or its behavior is senseless or even crazy. We like to congratulate ourselves when our own wise behavior saves us from an animal's viciousness. An excellent example of this is when people "narrowly escape" being "eaten" by a white shark. Most humans survive white shark "attacks"—this despite the fact this animal can weigh more than a ton, has a sensory system that can detect the electrical current of a beating heart, and possesses three hundred razor-sharp, serrated teeth capable of severing the head from a two-ton bull elephant seal. Do you think a person can escape such a predator? Not likely. People survive because the shark never intended to eat them in the first place.

A human on a surfboard may look like a seal—particularly when seen from below, where the shark is usually coming from. Most of the time, though, the shark instantly realizes the mistake and spits the person out.

Another error people make is concluding an animal does something for "no reason." Once, when I was in Borneo at Biruté Galdikas's orangutan study and rehabilitation camp, a volunteer who was smitten with the orange apes rushed up to one particular female whom she had met just the day before. She wanted to hug her. The orangutan promptly slammed the woman to the ground— "for no reason!" the woman said in hurt dismay. But the orangutan had a perfectly fine reason for her behavior: She didn't feel like being grabbed by a stranger.

During this same visit, Biruté's husband, Pak Bohap bin Jalan, told me through a translator that sun bears—small, shorthaired ursines that look like fat, short-legged rottweilers—were vicious and untrustworthy beasts. As proof, he explained that a sun bear had once, years earlier, attacked him "for no reason." How terrible! What was he doing, I asked, when he became the victim of such an unprovoked attack? The answer: He had been stuffing her cub into his shirt.

Even if it's not evident to us, animals have reasons for what they do, usually excellent ones. They have the same basic motives we do. They want to eat when they're hungry and sleep when they're tired. They love and protect their mates and their babies. They sometimes want company and other times want to be left alone.

We can't always assume that animals experience and react to the world exactly as we do. Otherwise, dogs would not eat horse manure. Fish would try to escape from the water. But in the ongoing practice of trying to understand what happens in an animal's head, there's a far worse mistake than assuming animals think like we do—and that's the crazy notion that animals' thoughts and motives are nothing like our own.

What is more interesting than the mistakes we can make when misinterpreting an animal's behavior is the fact that quite often—despite our very different bodies and different sensory systems—we understand each other very well indeed. Our views on many matters are often strikingly similar.

Consider a Stockholm University study of the aesthetic values of different human faces. The researchers presented undergraduates with photos of the faces of thirty-five young men and women and asked them to choose the most attractive ones. Then they asked a group of chickens the same question. The scientists trained chickens to peck at "average" human male and female faces. They then created images of faces with increasingly exaggerated male or female characteristics, and measured how often the chicken pecked at each. Which faces looked most like what the chickens thought a human face should look like? The chickens' preferences overlapped with those of the humans 98 percent of the time.

Why should this be so? Possibly both humans and chickens favor symmetry. Interestingly, it's well documented that chickens and people recognize each other by the same means: by looking at the face.

Fear of the Dark

— *Liz* —

*I*f your child is afraid of the dark, give the child a night-light. That child knows what he or she is doing, although perhaps not consciously. We are born with fear of the dark, although as we grow up most of us overcome it. But we can notice a residue of fear when we come home on a dark night. The first thing we do is turn on the lights. It's worth noticing the slight sense of relief we feel when the light goes on and we see what's around us. For this we can thank our ancestors.

We inherit this fear because it helped us survive. If a fear is hardwired, we don't have to learn it, which would have been risky, considering what it's about. And this, says ecologist Craig Packer, was probably lions.

In the fascinating paper "Fear of Darkness, the Full Moon and the Nocturnal Ecology of African Lions," Packer and his coauthors discuss fatal attacks on people by lions, which most often take place during the first few hours of the nights that follow the full moon.

Today we don't pay much attention to the moon, and not everyone knows that the full moon rises at sunset and on the following nights it rises later, getting smaller, until the last three days of the moon's cycle, when it rises with the dawn. We don't see its little crescent until late afternoon. After that, until it's full, it's in the sky before the sun sets, so it's already there when it gets dark. On those nights, our ancestors had less to worry about because they could see what was around them. And so could the animals that lions normally hunted, so when the moon was lighting the sky, the lions would grow increasingly hungry. All this changed on the night after the full moon—the nights were dark for increasingly long periods of time, and the lions were hungry. It was then that the people were in danger.

People of industrial societies no longer cope with lions, but people in African villages do. In one of the languages spoken by the Kalahari San—the first people, and also the last, to live as hunter-gatherers—a metaphor for "lion" is "moonless night." In the not-so-distant past, at least two such societies held their important safety-promoting dances in relationship to lions, the San dancing on nights of the full moon and the Hadza on the nights of absolute darkness. During the dance, the San trance dancers ran out in the dark away from the dance fire and cursed the lions, telling them to go. And according to Chris Knight, an anthropologist who studied the Hadza, the Hadza referred to the songs they sang during the dance as scaring away predators. They said, "We are singing for our lives."

When all of us lived as hunter-gatherers, the lions knew about our encampments and had no reason to fear us. Our weapons were spears with stone points—not much use against a lion—or perhaps poison arrows, which take days to kill a victim, or fire, which the savanna animals would have known as well as we did. A raging brushfire started by lightning might seem scary to a lion, but a campfire would not. If a lion came near one of our encampments, the firelight might show his shining eyes, and we could pull a burning branch out of the fire and shake it at him—that's what the San did—but many lions seem to know that their eyes shine and can approach without being noticed, especially if it's very dark. If we tried to run away, they'd catch us. The world's fastest person once ran at twenty-seven miles an hour, but almost any lion can run at thirty-seven miles an hour, at least for a short distance, and that's all she needs. We would have been easy prey.

So we would see the full moon with mixed emotions: It would light up the world all night that night, but its presence would signal that dark nights were coming when lions would be hungry. Thus our fear of the dark had survival value. It kept us near our campfires with our children close by. Surely we've had it since we came out of the trees, so it must be older than our species. Without it we might not be here.

Fear of Snakes

— *Sy* —

*I*n a single glance I could see thousands. They were all around me. When anyone moved I could feel it. Normally I dislike crowds, but happily I was not in a stuffy room at a busy party or a packed stadium. I was delighted to be sitting in a pit with 18,000 snakes.

The Narcisse Snake Dens in Manitoba, Canada, host the largest aggregation of snakes in the world. They're all red-sided garters—who, like nine-tenths of world's more than three thousand snake species, are harmless. Sitting among these soft, shiny, beautiful reptiles was thrilling.

But not everyone would like this. In reports from Gallup polls to YouGov surveys, snakes consistently top the list of Americans' fears. Snakes are ranked scarier than public speaking, heights, or even death.

Marion Lepzelter is used to seeing the terror. As a volunteer at the New England Aquarium, she used to handle the anacondas—constricting snakes who are strong enough to subdue and eat jaguars in their native South American jungles. Some of those anacondas were fourteen feet long. "How can you *like* snakes?" folks used to ask her. "How can you touch that thing?"

Members of the visiting public were astounded to see one of the giant snakes, a female named Ashley, slither up to the slender young woman sitting in the anaconda exhibit and place her head trustingly in Lepzelter's lap, while curling her tail around her legs in a serpentine embrace. Ashley clearly trusted Lepzelter and enjoyed her company; Lepzelter felt the same about the snake.

"They are beautiful. They're amazing. Every single one of them had their own personality," agrees Jennifer Berry, a librarian for the Pima County Public Library system in Tucson, who has owned pet snakes since she was eight.

"A lot of people think snakes are stupid automatons," says Berry. They're not, of course. Snakes learn to recognize individual humans, as anyone with a pet snake well knows. Some are bold; some are shy. Some like to cuddle. Some prefer to be left alone. "They'll communicate," Berry learned. "You just have to listen and watch. Animals will tell you what they're going to do. You can relate to them on a pure, honest level. They're far more complex than you give them credit for."

A special few Americans do appreciate snakes. According to the US Department of Agriculture, possibly as many as 2.5 million of us keep reptiles—mostly snakes and turtles—as pets. Snake keeping is even more popular in England. The Federation of British Herpetologists says there are now as many as 8 million reptiles and amphibians being kept as pets there—outnumbering the estimated British dog population of 6.5 million. (Of course, snakes and other reptiles and amphibians can be appreciated without welcoming them into your home, and they are almost always better off living in their natural habitat than in your house.)

But getting the good word out to the larger public about snakes is an uphill battle. Many people believe that humans are born with a fear of snakes. Experiments have shown otherwise. Babies don't fear snakes; it's a learned response. University of

Virginia researchers Vanessa LoBue and Judy DeLoache showed videos of snakes to babies, some as young as seven months old. The infants watched calmly.

In another experiment the same researchers paired silent videos of snakes with the sound of an adult speaking in frightened tones. That's when the babies got upset. Concerned, the babies watched the snake video closely. When the snake videos were paired with an adult speaking in an upbeat, happy voice, the babies paid no special attention to the snakes. They watched with the same level of calm interest with which they viewed other films of elephants, birds, polar bears, and other animals.

But the most interesting aspect of human fear of snakes is that while it is not inborn, it is extremely easy to learn.

Martin Seligman, a Pennsylvania psychologist now better known for his studies of happiness, was one of the first to show why these fears are easy to come by. We may be genetically preprogrammed, or "prepared," to learn to fear dangers that confronted our early ancestors. He found that while it is possible to make volunteers fear pictures of flowers or clouds by pairing the images with repeated electrical shocks, it took only two small jolts to induce a phobia of pictures of spiders or snakes.

Why might this be? Humans evolved in Africa, which, unlike North America and Europe, has a large number of venomous snakes, many of whom hunt primates. So it's no wonder that our close relatives also learn quickly to fear snakes, as Michael Cook and Susan Mineka showed in experiments published in 1989 with captive rhesus macaques. These monkeys had never seen a snake and didn't react fearfully to a snake-shaped toy.

But then researchers turned on the videos. One group of monkeys watched a fellow monkey acting frightened of a plastic flower. Another group saw a movie of a monkey scared by a plastic snake. The videos were edited to make sure the monkeys in the films looked equally upset. But the audience reacted quite differently to the two objects: When handed a flower, no one was afraid.

But when handed a toy snake, the monkeys who had watched the snake horror film were fearful.

Wild snakes apparently feel the same way about us. Most of the time, when snake and human meet, both flee in alarm. If we could only get to know each other better, things might be different.

Though gentle, the snakes I met at the Narcisse dens were not shy. They had just woken, chilled, from their winter slumbers packed deep into limestone sinks. Discovering the welcoming tunnels of my sleeves, they happily slid inside to warm themselves next to my skin. I was honored.

The Spirits among Us

— *Sy* —

*H*alloween pumpkins, Christmas wreaths, and Easter eggs summoning the spirit of the holidays help remind me to keep celebrating the ordinary spirits that animate our world.

In other cultures, spirits are on everyone's mind every day. While working on a book on moon bears, a now-endangered species once common in many countries in Asia, I traveled around Thailand, Laos, and Cambodia. Everywhere I'd find little structures atop short pedestals that looked at first like elaborate birdhouses. They're called spirit houses, and they fascinated me.

I found spirit houses in remote villages and at the edge of shopping center parking lots. They were outside restaurants, gas stations, and hotels. There was even one at a Bangkok internet café.

Some were quite well appointed. The one by a hotel I stayed at in Thailand, for instance, was flanked by two tiers of flower boxes and had a whole herd of stone and carved teak elephants massed outside the house proper. Another spirit house, outside a travel agency in Chiang Mai, had tiny Christmas tree lights strung all over its pagoda-like roof, which winked at night like fireflies. All the houses were stocked with rich offerings for the spirits: fresh rice, bananas, and sweets, prepared daily. Each morning I could see that the offerings had been consumed.

One night, on Jomtien Beach, Thailand, I crept out from my hotel to visit the spirit houses and see if anyone showed up.

I was not disappointed.

Bats flitted in and out of the flower boxes lining one spirit house, sipping the nectar. Mice and rats hurried to carry off grains of rice from a spirit house "porch." At another I saw a flying squirrel steal away with a slice of banana. And at nearly all the spirit houses, ants flowed up and down to partake of the food offerings.

In early morning, spirit houses provided a wonderful bird-watching opportunity. Birds would fly in, perch on the elaborate structure, and feast on the offered grain.

Did this mean the local people were duped into believing supernatural creatures took their offerings, when in reality ordinary animals were eating them? I don't think so. Southeast Asia is full of spirits—from river spirits to mountain spirits to (according to at least one hill tribe in Thailand) a spirit who presides over the cooking of tofu. But unlike gods, who are worshipped, and ghosts, who are feared, the spirits for whom these houses are built are rather ordinary. They're neighbors—neighbors whom nobody wants to offend.

The houses are built for the spirits who inhabit that particular place—like the one who lived in a tree that used to grow where a house now stands, or those who inhabited the soil now paved over for a gas station or parking lot. All of these spirits—from the ants to the birds to the squirrels, and surely many other creatures as well—had a perfectly good home until some human came along and took it. These structures provide alternative housing and food for the rightful owners of the place.

Spirits are not necessarily supernatural. The origin of the word *spirit* is the same as for the word *inspiration*—the common root is the Latin word *spiritus*, which means "breath." The spirits I met feeding from the spirit houses were all living, breathing creatures—and in many cases, creatures whose homes actually were usurped by a manmade structure, just as people believed.

Spirit houses remind us that resources and space are finite, and that in our hunger for more houses, more hotels, more shops, more gas stations, we humans continually disrupt and displace other beings. The least we can do for them is make some offerings to those who live among us, like neighbors—and who, like us, are hungry and eager to go home.

PART TWO

------◆------

Birds

*W*hen we think of commonality with animals, we think of common ancestors, which for us would be bonobos and chimps, or the gibbons before them, or the monkey types before the gibbons and maybe the lemur types before the monkeys. That would involve about 55 million years, so we seldom think back beyond lemurs. Yet here we are, finding commonality with birds, and we would have to think back more than 300 million years before finding that common ancestor—a lizardlike protoreptile descended from a salamander-type amphibian with scales, who had learned how to live on dry land.

So in evolutionary terms, a midsize ape known as a human is a long way from a tiny avian dinosaur known as a hummingbird, three inches long, weighing less than three grams, flying around that human's red-colored, syrup-filled hummingbird feeder. For two utterly different organisms who have been on totally different evolutionary pathways for 300 million years, isn't it strange to have a feeder in common?

The question may never be answered, but it's nice to think about—it suggests what works and what doesn't, so that organisms with vastly different characteristics can make sense of the same things, albeit for different reasons. As for the feeder, the bird is getting sugar to strengthen herself for the arduous migration she will soon be undertaking, and the midsize primate is gratifying social desires he inherited from his forebears over thousands of years, showing that often enough a cooperative social environment can be better than a selfish, combative one. In his case, to expand his social connections he bought a feeder to attract a hummingbird.

But that's minor compared with some other examples in this section, where you will see the story of a cockatoo who dances to music. The music has nothing whatever to do with birdsong, is a completely human invention that no bird in the wild has ever encountered, was totally alien to the protoreptile ancestor, and yet this cockatoo not only understood what it was for but unerringly danced to the beat.

Do we see why this is amazing? Songs serve a different purpose for birds than our music serves for us, so Snowball's ability to internalize and act upon a very complicated, multilayered noise that was produced entirely by an alien species could not have been predicted. When we look back 300 million years to the protoreptile who gave rise to generations of ever-changing life-forms, we struggle to imagine the thousands if not millions of evolutionary issues this must have involved, and when we see that one result is a bird dancing with a person, we're confounded. Our grasp of science, whatever it may be, doesn't help us here. All we can do is say, "Oh!"

What else was trundling down these lines of descendants? One characteristic appears again and again, not only in us but also in birds, which is the concept of doing things together. No doubt the practice was meant for same-species cooperation, but it's a good concept and often works well, and sometimes enough of its essence leaks out to cause one species to help or cooperate with another. Such are the puffin and his science-minded, caring followers, as shown in this section, also some fascinating chickens and their keepers.

Yet what we easily accept as normal should strike us as the most astonishing—our relationships with birds of prey. The unit formed by a person with a falcon on his glove could almost be the ancient, lizardlike, ancestral vertebrate put back together after 300 million years. A falconer and falcon are essentially one unit with a single purpose: The falcon gets a rabbit that both of them may eat.

Birds

Birds and people together . . . what to make of this? About all we have in common is the basic vertebrate body and the scales from our ancestral protoreptile's skin, which came to birds as feathers and to us as fingernails. We do share a few other tendencies, however, one of which is to notice elements of ourselves reflected in other species. The falcon who returns to the falconer, the hummingbird who alerts the human when the feeder is empty, the cockatoo who matches his steps to the human's while dancing so that both are on the beat, what do these tell us? Could the lizardlike ancestor, taking sips from a stream in eastern Pangaea 300 million years ago, have had genes that gave her a spark of affinity, perhaps with a fish she happened to notice? Were those genes in the embryos she left behind when she buried her eggs in leaf litter?

A question like this requires an important, inspiring answer, but here we're still asking. I like to think that in the future, scientists could discover some fossilized thoughts that fell out through that protoreptile's nostrils when she looked down at the fish. If they're in sedimentary rock and if radiometric dating suggests that they fell from their host approximately 300 million years ago, we could safely assume they belonged to a member of our ancestral species. We'd know more if such thoughts could be found and studied.

—Liz

Birds Rock the Beat

— *Sy* —

I've always loved to dance, whether in aerobics class or a friend's basement, in massive marathons or alone. But a partner makes dancing way more fun—especially when your partner is a parrot.

One of my favorite dance partners ever was a sulfur-crested cockatoo named Snowball.

"He *loves* to dance," his owner, Irena Schulz, told me when I phoned her to arrange a visit. "He's like the Energizer Bunny. He keeps going and going and going!" When I traveled to Indiana to meet him, I would experience Snowball's energy firsthand.

I'd first learned of Snowball on YouTube. From the moment it was posted in 2007, the little movie of the white parrot with the yellow crest dancing to "Everybody" by the Backstreet Boys went viral. Snowball came to Schulz as a rescue (she runs Bird Lovers Only parrot rescue in Duncan, South Carolina), but today he's a celebrity. Since his YouTube debut, he has danced on *The Tonight Show*, *The Late Show*, and *Good Morning America* and made many more videos.

If you watch Snowball dance, you can't help but smile. Bobbing his head, throwing his crest, high-stepping with his feet, the parrot is clearly having a blast. But he doesn't just enjoy dancing: He is *good* at it. Thing one to get right with dancing—far more important than grace or style or originality—is syncing with the beat. (Ever dance with someone who couldn't do this? Awful!) This is what most interested me about Snowball's dancing, because it bespeaks a quality of mind that most researchers had assumed belonged to humans alone.

Perceiving rhythm is an act so natural to us humans that we tap our feet to music without realizing it. But it's actually a sophisticated cognitive feat with a fancy scientific name: beat perception and synchronization. And it's similar in many ways to skills we employ when we use language. After all, as Tufts University Professor of Psychology Aniruddh Patel points out in his book, *Music, Language and the Brain,* both music and language are composed of strings of organized sound, full of meaning for both performer and audience. Both powerfully affect emotions. Both have richly structured patterns of rhythm.

For our first dance I selected a favorite song I hoped Snowball would like: "The Lion Sleeps Tonight" performed by the Tokens. The beat captured Snowball's interest immediately. Snowball stepped onto Schulz's hand for the short ride to the back of the gray swivel chair that is his favorite dancing platform and brings him roughly eye level with his people. His yellow crest rose high; his dark brown eyes shone with excitement. He bobbed his head enthusiastically as he first raised one leg, then the other, in time to the beat. Schulz and I joined in, dancing a version of the Pony. Because I was new, Snowball kept his eyes on me. Was he following my movements, I wondered? No—I am quite sure it was his idea to wave his left claw twice, then his right one once, while bobbing up and down and leaning slightly to one side. By the end of the second song we danced together ("Come Back to Me" by David Cook), I was certain: It was I who was following him.

How can I be so sure? Because Schulz's husband kindly video-taped that second song and emailed it to me. Watching a video can show you details you miss when you're experiencing the event real time—as Patel, the specialist on music and brain, well knows. Patel, too, came to watch Snowball dance, but it was the videos that yielded the most insights.

Patel videotaped Snowball dancing to the same song in different tempi. He then played the videos in sixty-frames-per second time resolution and then had them scored, with the sound off, by coders who didn't know which tempo was being tested. Colleagues also analyzed a video of birds moving to a beat (you'd be surprised how many dancing birds are on YouTube!), with the same results. His shocking conclusion: Birds are indeed capable of syncing to the beat of music, an ability previously thought to belong to humans alone.

This shows us that parrots, and probably all birds, "are hearing the world in a very complicated way, similar to the way we do it. It's suggestive of complex thinking," Patel says. "Only certain types of brains" can perceive, create, anticipate, and sync to the beat of musical patterns.

Ancient tales, in endless variation all over the world, claim that birds taught humans to talk, to sing, and to dance. The Fang-speaking people of Cameroon claim African grey parrots brought speech to people as a gift from God; the Kachin of Burma still reenact bird-taught dances today. Music and language may be considered glories of human culture, but their twined evolutionary roots run far deeper than our short human history.

For me, watching the video Schulz's husband had made of Snowball and me confirmed what I had felt at the time: Never before had I synchronized so effortlessly with a dance partner. I didn't know the second song—Schulz's spouse had picked it—so I expected my own dancing to be a little clumsy. It was not. We moved together as if in a mirror. And no wonder: I was taking my cues from a virtuoso.

Puffin M

— *Sy* —

"**O**h no!"

These words seldom signal good news, but coming from my stoic husband from his home office upstairs, I knew something was seriously wrong.

"What happened?" I called up to him.

"I just got this about Puffin M," he said, coming down the stairs to share his distress.

He had just opened the envelope from Audubon Project Puffin's seabird restoration program. "I am sorry to report that Puffin M did not nest on Egg Rock in 2015," read the letter. "He was seen ashore only once—atop the Razzo Rock Loafing Ledge on July 12, 2015."

Puffin M was missing and presumed dead.

Now we were both upset. I had never met Puffin M, yet I felt I knew him. My husband had "adopted" him for me as a birthday present starting in 2013. As a result, for the past two years I had received a photo of the handsome little black-and-white fellow with the distinctive clownish red-and-yellow beak, a biography of the bird, plus a detailed summary of his activities on Eastern Egg Rock, Maine.

A number of savvy conservation organizations, animal reha-bilitators, and zoos use animal "adoptions" like this to raise money to support their worthy causes. We also sponsor a twenty-three-year-old female Atlantic white whale named Calvin through the New England Aquarium. You can "adopt" a snow leopard through the Snow Leopard Trust; an orphaned orangutan through Orang-utan Foundation International; a shark through Shark Angels; and just about any other creature you can imagine. Just google "adopt a ___" and fill in the blank.

For me, adopting a puffin was the perfect present. I've been a puffin proponent ever since I first learned about Project Puffin's efforts to restore the birds to their historic nesting islands in the Gulf of Maine. Back in 1973 only two nesting colonies in Maine survived; the rest had been killed a century before. But that year ornithologist Stephen Kress came up with the maverick idea to transplant baby puffins (adorably called pufflings) from Canada to Eastern Egg Rock off Maine in hopes that, years later, they would return to nest there. Enticed by decoys and audio recordings, the first four pairs returned to nest in 1981; now a thousand puffins raise their young on five Maine islands.

I'd tried to get out to see the puffins but was foiled by bad weather. But now, thanks to Puffin M, at last I had my very "own" puffin to follow—even if I couldn't visit him.

Because of the long-term research on Eastern Egg Rock, I knew more about Puffin M's genealogy than my own grandparents'. M's parents, Y54 and MR314, had one of those May–September romances: Y54 was twenty-three years old; MR314, only six. But it was a successful pairing. When Puffin M hatched on June 20, 2000, in Burrow Number Three on the south end of the island, he became what would be the eldest of nine pufflings his parents raised over their eight years together.

Puffin M himself was a late bloomer. He didn't breed until age seven—two years later than normal. That was in 2007. It took seven attempts before he and his mate raised a puffling. But when it finally happened, I was able to read about it in more detail than a gossip magazine could dish.

"Researchers first spotted M on June 3 at the O Burrow entrance in the Big Boulder Jumble area on Egg Rock," I would learn.

On July 4 an unbanded female, M's mate, was seen delivering food to the burrow where the couple was raising their single puffling.

"On July 25, M delivered a single large herring to the burrow. . . ." Why do details like this so delight us? Why is it more fun to adopt a particular, named animal than to simply make a donation for conservation of a species?

Because, as researchers confirm, no matter the species studied—be it puffins or snow leopards or even sharks—animals are individuals, just like we are. And their stories can be as compelling as our own.

It's through stories about individuals that we make lasting connections. Ideas interest us; individuals move us. It's hard to imagine saving a whole species. But helping one individual? That's something anyone can do.

Puffin M's disappearance was surely even more upsetting news to the researchers at Project Puffin than it was to my husband and me. They knew him since he hatched. He mattered to them, not only as a precious and unique individual but as a member of a colony that staged a hard-won comeback—a comeback now facing new threats: Besides invasive plants and human disturbance, global climate change can profoundly affect the abundance of tiny herring and hake that pufflings need to survive.

So along with the bad news in the letter came a plea, signed by Stephen Kress himself: "We hope you will continue to support our efforts by letting us assign a new puffin," he wrote.

You bet we will—and for me, that'll guarantee another very happy birthday.

Chicken Indestructible

— *Sy* —

O n one of the coldest, darkest nights of January, our oldest chicken did not come home.

During all four seasons our flock of hens, collectively known as "the Ladies," range freely over our property (and that of our next-door neighbors—who for years kept cracked corn in their shed to feed our hens by hand). As evening approaches, the Ladies always return to the barn. Usually by sundown they're sitting on their perches, waiting for me to deliver their dinner of grain and scraps, stroke their feathers, and close the door to their coop, protecting them from night predators.

But on that freezing Friday night, my senior hen was missing. A member of the heritage Dominique breed—the ancestor of the Barred Rock—she was the last of a clutch of chicks who had arrived at our house, still egg-shaped fluff just a day old, more than eight years earlier. The average life of a pet chicken is only five or six years. (Though one I read about, a Red Pyle bantam named Matilda, reportedly lived to sixteen.) The Old Lady was a survivor, and while I love all our hens, I especially admired her.

Being a free-range chicken is risky business in New Hampshire. This past fall we'd lost another Dominique to a goshawk. In previous years other predators on our flock have included skunk, fox, mink, ermine, and neighbors' dogs. The Old Lady, in fact, knew what it was like to be in the mouth of a fox. On a summer day five years ago, alerted by the flock's distress calls, I had raced to her aid, screaming at the fox, who dropped her at the end of a Milky Way of white and black feathers. Eventually her bite wounds healed, her feathers grew back, but a broken, twisted toe remained forever after a testament to her brush with death.

Lately she had been showing her age. She was menopausal. Her crippled toe pained her, and she walked slower than the others. And though she still relished hunting for bugs and dust bathing in the barn's dirt floor, she had started sleeping in one of the nest boxes rather than perching next to her best friend, a bossy Black Australorp who pecked at all the spring chickens until they ran away squawking.

I called and looked for her everywhere that night but eventually gave up. Reluctantly I closed the small coop door to keep the other Ladies safe and warm, but I left the larger barn door open and the light on should she return.

Sometimes a hen will spend a night out on her own. Perhaps she was napping elsewhere when I closed them in; perhaps she saw a predator and was hiding in thick bush. Usually the straggler is back by morning. But the Old Lady was not.

Then I got the news of the wildlife sightings on our street. A friend a few houses down reported that she had seen, in her yard, two fishers—large weasels strong enough to kill house cats (though analysis of stomach contents has shown they do this less often than people think). The same day, she saw a goshawk and an owl—fighting! Later that weekend the same neighbor saw a bobcat in her driveway. "Now I know what happened to my missing chicken," I wrote on her Facebook page.

I knew our elderly hen was done for. She was ancient; temperatures each night had dropped way below zero; and the neighborhood was crawling with predators. It was pointless to even look for a carcass. I kept the other Ladies closed in.

And then on Monday afternoon, when I went to bring the cooped-up flock some arugula and cottage cheese to alleviate their boredom, there she was: None the worse for wear, the Old Lady was standing right in front of the coop door, clucking loudly.

How had she survived?

There are many answers to this question, but 8,325 of them are surely feathers. (That's the number of feathers a very patient person counted on a Plymouth Rock hen.) Made of the same protein, keratin, as fingernails, hair, horns, and hooves, these elaborate, heat-trapping structures are "the lightest, most efficient insulation ever discovered," writes conservation biologist Thor Hanson in his book *Feathers: The Evolution of a Natural Miracle.* The average songbird has 4,000 of them; the tundra swan, 25,000. Feathers are the reason even the smallest, most delicate-looking songbirds don't freeze in the winter. Biologist Bernd Heinrich, professor emeritus at the University of Vermont, once calculated the difference between the temperature beneath the feathers of a tiny North Woods bird called a golden-crowned kinglet and the subzero outdoor nighttime air: The bird can be 140°F warmer! No wonder we make our best parkas and comforters from bird down.

But there's another reason, I think, that the Old Lady survived that long freezing weekend among a gauntlet of predators. And that is that the Old Lady is smart.

Chickens in general are smarter than most people realize. An individual chicken can recognize and remember more than a hundred different chicken (and presumably human) faces; experiments confirm they have a great affinity for spatial learning, even in the absence of obvious landmarks. But each chicken is an individual, and some are smarter than others.

The Old Lady, I think, was one of the smartest. She had clearly outwitted me, a tool-using primate with a brain that outweighed her entire body. It had snowed lightly every day that weekend and also the day she returned. Though I had looked for her extensively, only later did I figure out, thanks to the lack of chicken tracks in the freshly fallen snow, where she'd been that weekend: tucked safely into an old pile of hay in the back of the barn—where she left me one giant egg, frozen solid.

Hawk Migration

— *Sy* —

My understanding of raptors changed forever on the day a hawk first landed on my fist.

Her name was Jazz. She was a four-year-old Harris's hawk, a species native to the desert Southwest, but she was living at the New Hampshire School of Falconry in Deering, where I was taking a lesson. At the toot of a whistle, the powerful bird of prey flew toward my outstretched arm. Her wings, spanning nearly four feet, blew back my hair at her approach. It made my heart sing.

Smack! Her huge yellow feet and ebony talons gripped my leather-gloved hand with shocking strength. Then she began the work for which her kind is named. Birds of prey—raptors like Jazz—are meat eaters, and the reason she had flown to me was to eat the piece of cut-up partridge that was waiting for her there. Her piercing, mahogany eyes focused on the job of tearing into the flesh with her sharp beak and feet with an intensity stronger than rage and brighter than joy. Inches away from my face, I beheld in this bird of prey a pure wildness more blindingly alive than I had ever before seen.

But here's the most amazing thing: Each fall and spring, not far from your house, thousands of birds of prey of dozens of different species may be flying over your head.

Where we live in New England, the fall migration of day-flying raptors—hawks, eagles, peregrine falcons, kestrels, kites—is one of our great wildlife spectacles. Yet most people never notice them. And fewer still realize who these birds really are.

They are tigers of the air.

Though many birds hunt—robins eat worms all summer, after all—raptors are the only birds who are exclusively predatory. They are such good hunters that they used to hunt and eat our ancestors. Recently, reexamination of a famous fossil hominid, known as the Taung Child, discovered in South Africa in 1924, concluded that the three-year-old was not killed by a leopard as previously thought but by an ancient relative of the crowned hawk-eagle. (The modern hawk-eagle still hunts monkeys in the same way.)

Wild raptors don't hunt children today, and you're in no danger. But to capture prey that may be large, fast, and smart, birds of prey employ powers that should leave us in awe.

They literally see the world in a different way than we do.

All birds need excellent eyesight to fly; but in birds of prey, who hunt on the wing, the sense of sight is developed to a super-power. An eagle flying at a thousand feet can spot a rabbit across a distance of nearly three square miles.

In birds of prey, the eyes weigh more than the brain; they have better distance perception than other birds. With forward-facing eyes, they have binocular vision like ours, only better. Fields of view of the left and right eye overlap—which allows the raptor brain to calculate distance instantly by comparing the different images from each eye.

Such accurate eyesight is essential when you might be diving upon your prey, as does a peregrine falcon (the fastest bird on earth), at more than two hundred miles per hour. But during the autumn, raptors are displaying not only speed but also endurance.

Migrating eagles and hawks may cover more than two hundred miles a day on their autumn journey to Central and South America.

It's a feat that demands some of them store as much as 10 to 20 percent of their body weight in fat before they undertake the migration. They conserve their fuel wisely. Migrating raptors soar on winds deflected up and over hills and mountains. They ride high on rising currents of warm air called thermals. Dozens to hundreds of raptors of different species may gather to take advantage of a good thermal in large, swirling aggregations called kettles. These kettles of hawks are amazing to see—but it's easy, if you're patient and know where to go.

There may be a hawk-watching site near you. (Daily counts at each site are posted at https://hawkcount.org, the database managed by the Hawk Migration Association of North America.) Each September my husband and I visit New Hampshire Audubon's raptor observatory atop Pack Monadnock in Miller State Park in Peterborough and join the crew of hawk-watchers there. On one Saturday, an hour went by and we saw only one kestrel. But on the following Monday, hawk-watchers counted more than 3,000 broad-winged hawks alone, and another 1,858 broad-wingeds were counted on Thursday.

Without a spotting scope, most of them looked like mere specks. But we knew what those specks really meant: thousands of tigers flying over our heads.

Bubbles Wrapped
in Feathers

— *Sy* —

*T*hey flash in front of flowers and feeders for seconds, wings a blur, and then whiz away. Next they're back— but before you can gasp at the beauty, they're off again. A glittering fragment of a rainbow; a flaming comet; a living gem: All of these metaphors struggle to describe the evanescent magic of hummingbirds.

But what they are doing when we don't see them is more wondrous yet—as I discovered several years ago. Working with a licensed wild-life rehabilitator, Brenda Sherburn, one summer I was privileged to help to feed, raise, and release orphaned baby hummingbirds.

Too often, people "rescue" baby hummers prematurely, Brenda told me. It's rare to find a hummingbird nest, but if you do, back off, leave the babies alone, and, using binoculars to watch from a safe distance, observe the nest without looking away for at least twenty minutes. "So few people can just sit still and watch anything that long," said Brenda. But if you so much as blink, you could miss the mother's return. A mother hummingbird leaves the nest from 10 to 110 times a day to find food for her nestlings.

To survive, a hummingbird must consume the greatest amount of food per body weight of any vertebrate animal. A single bird may drink its own weight in a single visit to your feeder—and seconds later come back for more. That's because a hummer breathes 250 times a minute. The resting heartbeat is 500 beats per minute, and the heart can rev to 1,500 a minute in flight. A film I watched claimed that a person as active as a hummingbird would need to consume 155,000 calories a day—and the body temperature would rise to 700°F and ignite!

An adult hummer visits an average of 1,500 flowers in a day. If the nectar were converted to a human equivalent, that would be fifteen gallons a day. But few people realize that insects are equally essential. Each hummingbird needs to catch and eat six to seven hundred bugs a day. (So spraying insecticide in your yard is like hiring a hummingbird exterminator.)

The food requirements mentioned above are for a single hummingbird. A mother caring for nestlings (there are usually two) needs even more. Lucky for us, Brenda had access to a fine compost pile with plenty of fruit flies, and her husband, Russ, was willing to catch fresh ones for us every day.

Each morning, when normal people were grinding coffee beans, Brenda would take out her mortar and pestle to grind flash-frozen fruit flies. Then she'd mix them with nectar, vitamins, enzymes, and oils. Because this food spoils easily, we had to make a fresh batch several times a day. From dawn to dusk we would deliver this to the babies' gaping beaks—by syringe—every twenty minutes.

Brenda was one of a handful of specially trained and deeply committed wildlife rehabilitators qualified to do this. I was honored to help. But for these fragile nestlings, each moment was fraught with danger. Miss a feeding and the babies could starve. Worse, explained Brenda, was what could happen if you fed them too much. "They can actually pop," she told me.

Hummingbirds are little more than bubbles wrapped in feathers. Our bodies are filled with organs; theirs are full of air sacs. Their feathers weigh more than their skeletons, and both their bones and their feathers are hollow. It's hard to imagine anything more fragile.

And yet our fragile orphans, like the hummers at your feeder, were born to conquer the sky. Brenda lives in California, which boasts several species; as their feathers grew in, our babies revealed they were Allen's hummingbirds. To impress a female, a male Allen's performs a plunging flight that makes it the fastest bird for its size in the world. In terms of body lengths per second, it even bests the space shuttle!

On the East Coast we have only the ruby-throated hummingbird, named for the flaming red throat patch on the males. These birds are equally spunky: Each fall they undertake a punishing migration across the Gulf of Mexico, which may demand twenty-one hours of nonstop flight.

It's shocking to realize that someone who hatches out of an egg the size of a navy bean is capable of such feats. But equally shocking is the gauntlet of dangers a hummingbird may face on an average day. Hawks, jays, squirrels, crows, even dragonflies eat them. They tangle in spiderwebs searching for insects (they also use the silk in their nests, to give them stretch as the nestlings grow). They fly into our windows; they're hit by our cars; they're poisoned by our pollutants. The most common reason for any bird's admittance to wildlife rehab is also our fault. It's abbreviated on forms as CBC: caught by cat.

And yet we can help. Put out a feeder. Plant nectar-rich flowers. Keep a compost pile. Support a wildlife rehab center.

Reasoning that surely a bird so tiny with feathers so brilliant must be born anew each day, the Spaniards who first encountered South America's hummingbirds called them "resurrection birds." This names the gift these birds offer us this summer, with each fleeting glimpse. They force us to see the world made new each time, and teach us to believe in ordinary miracles.

PART THREE

Dogs and Cats

*H*ow did we acquire dogs and cats? We didn't. They acquired us. Early on they would have seen the benefit of relationships with our species, not knowing that we would credit ourselves for these new relationships. For instance, it's often said that dogs developed from orphaned wolf puppies whom people adopted and raised. But back when we lived in the natural world, behavioral changes mostly resulted from the need for nourishment, never from the need of more mouths to feed. If the people who recently were hunter-gatherers—Namibia's pre-contact San or Bushmen—serve as an example, they did not consider other species as companionable, and they expected the dogs they eventually acquired to feed themselves. Since there aren't many ways to live by hunting and gathering, the culture of the people who joined with wolves to make dogs would in ways have resembled the culture of the pre-contact San, and if such people found an orphaned wolf pup, they'd kill it and make a hat out of the skin.

As early as 100,000 years ago, human hunter-gatherers and a now-extinct species of gray wolves lived in northern Eurasia where winters were long and the meat of large game was an important part of the diet. Often in winter, when vegetable foods were unavailable, meat could be the only food the people had.

They hunted the same game as the wolves, sometimes at the same time in the same places, as both would know when and where their game would be found. Wolves would have been quick to notice when the human hunters killed an animal such as a rein-deer—the hunters would remove the guts, the hooves, and perhaps the head, and would cut up the meat to carry to their lodge or encampment. After they left, the wolves would move in to eat the bloody snow and the body parts the hunters had discarded.

Both species hunted in groups and thus knew the value of cooperation, and both would have known about herding. Wolves are endurance hunters (unlike cats and people, who are stealth hunters), and perhaps wolves would sometimes drive a reindeer toward some human hunters, perhaps accidentally but possibly intentionally. A wolf could run a reindeer to exhaustion, but wolves don't always kill as the big cats do, by leaping on the victim with a choking bite to the neck. Wolves can bite the victim's neck, but also they take bites from the victim as they run beside her; thus one wolf alone might have difficulty making the kill. People with spears had less difficulty. Wolves could have seen an advantage here. Perhaps the people did, too. It might have led to cooperative hunting.

The next step may also have been taken by wolves. Fossilized wolf scats were found in human encampments, suggesting that wolves either were living with the people or came to scavenge when the people were away. The best guess here may be that they lived at the peripheries of the encampments, where they scavenged when they could, as do other animals in some areas.

Here I look to Marcus Baynes-Rock, an anthropologist who studies not only people but also the animals with whom they associate, and who describes what one might see as an ideal situation for future hyena domestication in his wonderful book, *Among the Bone Eaters*. In Ethiopia clans of hyenas live near certain villages where they are sometimes fed as a tourist attraction. They hunt as do other hyenas, but they also roam the village paths at night, sometimes finding something to scavenge, and they seldom if ever bother the people. A somewhat similar situation can be found among the dogs of many villages in the developing world.

These dogs are never fed by the people and have no food except what they scavenge. If thousands of years ago wolves scavenged those northern encampments without doing serious harm, the people probably tolerated them, especially if the wolves helped with hunting and alerted them to predators such as the oversize lions, *Panthera leo spelaea*, who inhabited the area at the time and probably ate wolves as well as people. Tolerance could have led to interdependence, which could have led to friendship, which could have led to dogs—at first to dogs who were very much like wolves, later to dogs with minor physical modifications due to interbreeding, and later still to different breeds where the breeding was controlled by humans.

It's hard not to like a dog. A doglike fossil from 33,000 years ago was found in a grave in Germany beside the fossil of a man. More such burials were discovered, these dating from 14,000 years ago, so the practice went on.

Some of us continue the custom. Our family favors cremation, so after my death I'll be ashes. All my dogs except those with me now are already ashes. When our ashes are mixed, we will all be together; thus I'm not so different from the guy in Germany who 33,000 years ago was buried with a dog.

Cats have a similar history, also food-related. Cats descend from a small African wildcat, *Felis silvestris*, who lived in what's known as the Fertile Crescent, the area that curved above the Red Sea from Egypt to Israel. It was here that people domesticated grass to make grain. Rats and mice ate the seeds of wild grasses, and people harvested the seeds as grain and stored them in their granaries. It can't have taken the mice and rats more than a few days to find a seed-filled granary and not much longer for the little wildcats to find the mice and rats.

For controlling the mouse and rat populations, the cats were greatly appreciated. They did no important damage and, unlike dogs, they didn't steal or even scavenge for the people's foods. As for living in a sheltered place where hawks and eagles couldn't catch them, and where grain attracted mice and rats, a cat couldn't ask for more. Cats and people made such a good fit that they became venerated in Egypt, but that was just two thousand years ago, and the trend began earlier. The fossil of a cat from eight thousand years ago was found on a Near Eastern island that could only be reached in a boat. The fossil looks like a wildcat's but was probably a pet cat's. Whoever rowed the boat had brought his cat along.

In my house and perhaps in yours, too, we find a mini version of the Near Eastern ecosystem. Grain was distributed by ships worldwide, and the local mice, rats, and cats went with it. Our cats are modern versions of *Felis silvestris*, and the little gray mice (not the indigenous white-footed deer mice) who live in our houses, as well as the so-called brown rats or Norway rats (they're not from Norway), descend from those Near Eastern ancestors, and the Near Eastern grain, by now somewhat modified, is the flour we use to make bread.

—*Liz*

Pets with Disabilities

— *Sy* —

*W*e weren't looking for a puppy. We were still grieving for Sally—the rescued border collie who had been the sweet, funny, demanding center of our family for nine years, who had died of a brain tumor. But then we got a call from our vet. He had just finished his exam of a litter of border collies from a nationally respected breeder, an admired acquaintance, whose pups famously grow up to work as professional herders. One of them had a blind eye. Would we take him?

We named him Thurber, after James Thurber, the iconic *New Yorker* cartoonist and essay writer who loved to draw dogs and who also had a blind eye. (His brother shot his eye out with an arrow during a game of William Tell.)

Our Thurber is cheerful, handsome, smart, and eager to please. The minute we brought him home, he filled my bottomless sorrow with endless elation. Folks tell us he's the cutest puppy they've ever seen. We forget he has a blind eye.

And one blind eye isn't much of a disability (unless you're an "eye dog" like a working border collie, who must move dozens, sometimes hundreds, of other animals by the force of your stare). But what often seems remarkable to us humans is that other animals who acquire disabilities later in life—losing a leg, an eye, their hearing—seem to get on with life just as joyously as Thurber has started his.

At an animal sanctuary in Lanzenhainer, Germany, half a dozen paralyzed dogs—using wheel carts in place of their back legs—play fetch in a field with as many able-bodied dogs . . . and one woman who's getting plenty of upper-body exercise throwing the stick. The woman, Gritta Goetz, is the director of the sanctuary, and she's hosted as many as eight paralyzed dogs at a time. Thanks to their wheel carts, she says, the disabled dogs "are fast enough to get [the stick]. And the smallest paralyzed dog is the one who can protect the stick against anybody," she says. "If she gets it, even the leader has no chance anymore."

These dogs clearly love their lives. But as well-known veterinarian Mark Pokras told me, some people automatically insist that an animal with a disability is better off put down. An associate professor at Tufts, he also ran the university's Cummings School of Veterinary Medicine's wildlife clinic for more than a decade, and he's had many spirited discussions with his medical colleagues, arguing for an animal's life. He knows firsthand that life can continue undiminished—or even enhanced—after a seemingly catastrophic illness or injury. He lost a leg to cancer in his twenties and it never stopped him from doing what he loves. If anything, it may have increased his compassion for his patients.

Though some people may feel sorry for animals with disabilities, the animals don't feel sorry for themselves. (Animals "do not sweat and whine about their condition," Walt Whitman correctly understood; nor do they "lie awake in the dark and weep for their sins.") Take Faith, the dog born with deformed forelegs, who learned to walk and run on her two hind legs in the posture of a person. Faith wowed Oprah when she appeared on her show in 2008. "She's a demonstration of what it looks like to persevere," her owner told the audience.

Animals like Faith don't need our pity. They have much to teach us.

Thurber is not the only one-eyed pup who lives on our street. Down the road August, a golden retriever, was born with such severe glaucoma that one of her eyes had to be removed. The puppy accompanies her owner, a psychologist who specializes in trauma, to the office. Of course, the clients love having a puppy around. But as August's owner tells me, the folks seeking help for their problems also love being around "someone else who isn't perfect."

As for our Thurber, every once in a while, when the light is right, we can see a greenish cast to his right eye to remind us it's blind. But I never think of it as a "bad" eye. It's his beautiful, blessed, beloved eye—the eye that brought him to us, and changed my sorrow to joy.

Breeding Dogs

— Liz —

hy do cats live longer than dogs? And why are most cats free of the bone deformities, respiratory problems, and other serious deficiencies found in dogs? The answer is that cats choose their own mates. Or most of them did until it became important to spay or neuter cats, partly because the shelters became overcrowded with unwanted cats who had to be euthanized to make space for more unwanted cats, and partly because the feral cat populations were mostly successful, thus too many cats were outdoors killing birds.

Choosing one's mate is the best way of breeding by far, and mostly it's up to the females. Yes, males come sniffing around as interested suitors, but females choose whom to accept. We humans do this, dogs once did it, and cats still do it if they can, choosing strength, good health, and high status.

Dog breeders have different agendas, encouraged by the various kennel clubs and dog-show providers to produce animals who are grossly distorted and seldom live as long as dogs bred for other purposes. As for me, I had dogs who were not spayed or neutered, and two of them, both Siberian huskies, produced three puppies so fit and healthy that as young adults they won just about everything there was to win in the New Hampshire three-dog races. Other contenders hated to see them coming.

All my huskies were fabulous as racing dogs. We have an attic full of trophies to prove it. They lived in good health to old ages, and if they'd had to live in the woods, they'd have made it on their own without a problem.

I'm not saying that every dog should be physically capable of living in the wild, although that's not a bad thing. I'm just saying they're better off without respiratory problems or hip dysplasia. And why do they have such problems? As has been said, they're bred to meet certain standards that allegedly include good behavior but overwhelmingly are for various forms of unnatural appearance as determined not by other dogs but by kennel clubs—here in the United States it's the American Kennel Club (AKC)—that issue the standards to breeders. For all the lip service offered by these kennel clubs about behavior, we ignore the fact that for dog breeds, behavioral similarities are minimal and individual differences are great, and when we accept the kennel club standards we say good-bye to strong dogs with good health and high status.

I adopted a small dog from a shelter. Supposedly purebred, he had AKC papers. We knew he was born on a puppy farm and was sold to a chain store such as PetSmart, then sold again to someone in a city apartment, so he left his mother far too soon and didn't know where to relieve himself or what grass was, or so it seemed, as he buried his nose in grass whenever he came upon it. He was looking for traces of other dogs, I later learned, but at the time I only knew that he was lonely and frightened. I was handed his AKC papers, but not wanting to be humiliated if anyone saw them, I tore them up and burned them in the woodstove as soon as I got home.

If not for what I had seen a few months earlier, I would merely have dropped his papers in the recycling bin when I got around to it. But before adopting this dog, I had watched an AKC dog show. One contender was a little Pekinese—a breed developed for fluffy fur and short faces—and as the poor little thing was paced around the ring in the supposedly triumphant final cycle, she was panting so fast and so hard in her desperate struggle for oxygen that she had to stop and walk. She would have liked to sit down to catch her breath, but she was forced to continue.

Dogs with flat faces have the same amount of respiratory tubing that dogs with normal faces do, but it's so scrunched up in their distorted skulls that air can hardly pass through. But to the judges, her condition was no problem. Of all the dogs present, and there were plenty of them, thanks to her fluffy hair and flattened face, this poor little gasping creature won Best in Show.

When humans get involved with other species, especially if they try to manage another species, things often get messed up. Ever since a dog's appearance made a difference, breeders have been manipulating what dogs look like no matter what this does to health and longevity.

The current solution to overpopulation by dogs and cats—overpopulation meaning that too many people have pets they don't want so they leave them by the side of the road or abandon them at overburdened shelters—has produced semi-rules for ordinary people (in other words, nonbreeders) to spay or neuter their pets, and while there's an argument for spaying females, there isn't for neutering males. Females are fertile for a few days twice a year but males are always fertile. Neutering disturbs their growth if done too early, as it usually is, and alters their personalities, so they never reach the kind of maturity that intact males do. It may be necessary to spay most females, but where is the reason for neutering males other than the politically correct theory that males and females should be treated equally? It's said that the more animals who are sterilized, the fewer unwanted pups will be born. Since modern dogs in developed countries seldom if ever get to choose their own mates, there may be a point here, but since it's the female who has the pups, this reasoning doesn't make as much sense to me as it may to some.

Another reason was offered by a veterinarian who announced that castration was necessary to prevented certain health problems. I wondered if he is castrated. If he's intact, can we believe him?

Tiny Dogs

— *Liz* —

To my surprise, after knowing medium-size dogs for more than fifty years, I found myself the owner of two tiny young dogs, a Chihuahua and a pug type. They came from a city, and I got them from a shelter. My husband was a historian whose specialty was Central Europe, and the dogs are named for famous Czechoslovakian authors—Kafka for Franz Kafka and Čapek (pronounced Chapek) for Karel Čapek, an ethnic Czech who wrote, among other things, *War with the Newts* and *R. U. R.*, a play that gave us the word *robot*. Kafka is the name of the bigger dog, the pug type, because in world literature, Franz Kafka is more important, thus bigger, than Karel Čapek.

With such distinguished names, these dogs became distinguished dogs who taught me things I'd never dreamed of. Among such revelations is the way they experience the world. They're so small and everything else is so big that they like to sit on people's laps. No wonder they're called lapdogs. Up on your lap they not only see you almost face-to-face instead of looking up at you as you might look at a treetop, but they can also see at a distance, just as the rest of us do.

Tiny Dogs

They're also charming and adorable. They sleep on your bed at night, pressed against you, sometimes on top of the covers but often under the covers with you—or that's what they do on cold nights.

I once saw Jane Goodall interviewed on TV. She didn't seem overly impressed with the interviewer, and when he asked what animal she liked best, her face remained blank as if she hadn't heard him. The silence became embarrassing, so the interviewer leaned forward. "Is it chimpanzees?" he asked eagerly. But Goodall's expression never changed. "Lapdogs," she said.

And I know why. Yes, they're charming, but they're also smart and interesting, and being too short to see at a distance, they rely on their sense of smell. We live in the country, with all kinds of odors wafting from the woods, but what interests these dogs is cars and grass. My country-born dogs had no such interests, so this surprised me.

It shouldn't have. My new little dogs had spent much of their early lives at the end of a leash walking down sidewalks. The grass along the sidewalks held the marks of other dogs. Grass was to them is what a list of names is to us. But why did they focus on cars?

At last it struck me that it wasn't the cars themselves that drew them, it was the tires. Not a car came in our driveway that these dogs didn't rush to smell the tires, usually but not always starting on the right—the side they would have sniffed if the car was parked on a street. They then would circle the vehicle, sniffing each tire carefully, returning to sniff the others again, and when they'd examined each tire to their liking, they'd mark some of them, lifting a leg.

Cars are the internet of dogs, and tires are their social media. From tires they learn of dogs far away, dogs they don't know and may never meet but will have left messages. These are sometimes called pee-mail, but such a mark is more like a tweet or a blog—one of your thoughts that you share with the world in general with no expectation of a reply.

So I've learned two things from lapdogs. First, like most dogs, their early years are the most important as far as their values are concerned. My house is in the country, surrounded by dozens of wild animals, but these don't interest my little dogs. And second, their social interests are fully as flexible as ours. By now they've had messages from Boston, New York, Chicago, southern Maryland, northern Virginia, and eastern Florida, and they've sent their own messages back, confirming what I've long suspected: that the most important thing to dogs is other dogs. Lapdogs prove this conclusively. Although mine live amid hundreds of acres of woods and fields, an unknown dog who lifted his leg in Boston seems more important than the local bears, deer, coyotes, porcupines, bobcats, skunks, or fishers. In this my dogs are like most people. If I were told of a local porcupine, skunk, or fisher, and also of a person I didn't know who lived in Boston, the person could seem more important to me.

Or that's what I'm saying. To be truthful, I'd rather know about the fisher, but I like to seem as orthodox as I can.

How Best to Educate a Dog

— Liz —

We humans don't learn our behavior from dogs, but dogs must learn theirs from humans. Humans and their elders live in a house where the elders and the youngsters are a family. They speak the same language, they understand one another, and the elders teach the youngsters what to do. But a dog brought to that house is a stranger. He can't understand the language. If he'd stayed with his parents and his siblings, he'd belong to a pack where he could learn from his elders, but in the new house he's a loner. Most dogs accept their owners as their elders and learn from them as best they can, but often it's a struggle. Does a cow learn easily from horses?

I've lived with dogs for more than fifty years and found, not surprisingly, that they learn best from other dogs. During those fifty years I kept a dog culture going. They saw themselves as a pack, the young dogs learned from their elders, and the results, including their housebreaking skills, were perfection.

The science of all this is swirling in my head. My story will prove that our best trainers are our own kind. It's not that I didn't train my dogs, but I'd train one at a time and only for fun. I had a little tool that made a click, and when the dog did something special I'd snap the clicker and give the dog a treat.

They loved it. They'd beg to do it, they'd leap around with excitement when we'd start, and they'd joyfully try all kinds of things—bark, jump, lie down, roll over, whirl around on hind legs, and so on—until they'd hear the click and get the treat. I'd say "sit" if the dog sat, or "roll over" if she rolled over, so she'd do the same thing again if asked, even without a treat, because she associated the process with such pleasure.

Clicker training is fun but really just a game and isn't much use for addressing negative behavior. To housebreak a dog by this method, one would need to catch the dog making a mistake, click the clicker, and instead of giving a reward, give him an electric shock. But when pack dogs learn from their elders, their housebreaking skills come easily. They understand what their elders want them to do, and they do it without even wondering about it.

For us, this pack training ended when the last dog of our fifty-year culture died. After pulling myself together I went to the local shelter and found the two little dogs I've mentioned in earlier essays, a small pug type named Kafka and little Chihuahua named Čapek. Kafka understood housebreaking but Čapek did not, and because these dogs were like siblings, not like a mature dog and a young dog, Čapek saw no need to copy Kafka when relieving himself. As far as he could tell, his pack and his elders were people and cats. All of them lived indoors, all of them relieved themselves indoors, and the obvious place to relieve oneself was in the house. Čapek would often pee near a cat box or a toilet.

Of course, I tried the recommended training methods—confining the dog or keeping him on a leash, taking him outside often and praising and rewarding success, but he cried if confined and didn't understand about the leash, and even after I gave up on these tools, keeping only the rewards, he still didn't really get the picture. Often enough I'd find a stain on the leg of a chair or on one of my boots in the closet, or notice him innocently lifting his leg against the umbrella stand as if he found this normal. He'd acknowledge his mistakes only if I saw him make them.

So I tried puppy pads: the canine version of incontinence pads used in hospitals. These worked well enough to some degree—he may have used them in his former home—and at night they were safer than a trip outside. A great horned owl who lives in our vicinity had floated silently out of the sky to capture and kill someone else's Chihuahua. I don't know how small that Chihuahua was, but mine is smaller than a cat. And he's black, so in the dark he's invisible to a human and to the owl he would look like a skunk—a common prey of large owls (which is why these owls sometimes smell bad). Even on a leash, which my Chihuahua didn't understand or like, he wasn't always visible.

Many months passed while he used the pads. But too often he'd miss and leave a mess on the floor. He meant well, but if such a thing happens and a human doesn't notice and clean up quickly, the house soon smells like a latrine. No wonder his first owner got rid of him.

But that may have been for the best. By this time many an owner might be punishing the dog severely, but I would never do that—the image of a shouting giant, ten times bigger than a Chihuahua, standing over him while he's helpless and disciplining him severely doesn't impress me. So remembering how well young dogs learned from mature dogs, I looked forward to a visit from my daughter, her husband, and their elderly Chihuahua, Boots, whose housebreaking was perfection.

Čapek thought the world of Boots. Wherever Boots went, Čapek would follow, noticing where Boots relieved himself even if Boots was just leaving a message, and Čapek would lift his leg to overmark, perhaps to remind Boots that he wasn't the only Chihuahua present, or perhaps to reinforce the message. By the time both dogs were back in the house, Čapek had nothing left to mark with, and I felt we were getting somewhere. After Boots and his family went home, Čapek rarely lifted his leg in the house until after the first snowfall.

He had never seen snow and was reluctant to go out in it. He weighed only nine pounds, his fur was thin, and his feet, which were the size of dimes, would freeze when he stepped in wet snow. I bought him clothes and boots, but he hated them and would run away and hide if he saw them. Thus he still saw the house as an acceptable winter bathroom, no matter how hard I scrubbed or how widely I sprayed deodorizer, and would acknowledge his mistakes only if I saw him make them.

With time, things improved. He'd learned something from Boots, and praise, rewards, and attention reinforced this. He still makes mistakes but not often, and by now I consider him trained, or trained enough, if not as well as my earlier dogs who were trained by their wonderful elders.

How many hours went into his training? I can't count them. How many hours did my older dogs take to train younger dogs? None. They didn't even train them. They just did what they did and the younger dogs copied them, and all this in minutes, no puppy pads, no mistakes to clean up, everything easy and natural.

This can be considered a scientific fact, and I offer my habits as evidence. I was trained by my parents, my grandparents, and the nanny—those who would have been my pack members if we were dogs. Today when in a house, I use the bathroom and only the bathroom, then only the toilet, never the floor, and when I'm outdoors I never pee just anywhere but only in the woods when on a hike. Čapek now knows about peeing outdoors but doesn't always. I'd often been told that Chihuahuas are hard to train, but I knew this already, and it doesn't explain why my success has been spectacular and Čapek's has been moderate to mild.

I learned from conspecifics and he didn't. That's what explains it. People were not his conspecifics. Kafka and Boots were not the right conspecifics. The members of my pack were humans and so were his. That's the secret.

Cat Vandalism

— *Liz* —

I once saw a photo of an enormous mound of fluff on someone's lawn and learned it was a sofa belonging to a veterinarian whose cats had caused its condition. This didn't surprise me in the least because my house, while reasonably neat and clean, has signs of vandalism. The vandals were cats who sharpen their claws. But rather than having the cats declawed, a monstrous practice that removes the end of the toe and causes pain forever because cats walk on tiptoe, I manage their vandalism by accepting it. My cats are far more important to me than our sofa or our papers or our pillows or curtains or wallpaper or that beautiful china cup made by a famous pottery artist in Prague or the four-hundred-page manuscript I wanted to send to my agent. The destruction of the cup resulted not from clawing but from the desire of cats to jump into cupboards when the doors are carelessly left open and to sometimes push things out. Two of our three cats can jump from the floor into a high cupboard and land on their feet, disturbing nothing, so these cats seem to push things out on purpose, perhaps to watch them fall. The third cat used to do this, but he's aging.

As for the manuscript, its destruction was unintentional. While watching a bird through a window, an engrossed cat stood on my computer keyboard with his hind foot on the enter key until the manuscript was five thousand pages long. I would have had to delete the whole addition one page at a time, so I decided to start over. I had a copy.

Our sofa is leaking fluff at the seams, the screens in our windows have holes in them, we no longer have curtains because the cats tore them while trying to climb them, and our hallway is hung with shredded wallpaper from the baseboard up to the height of a cat's reach when he's on his hind feet. That's the downside. The upside is that higher on the wall, the paper is smoothly in place, showing charming scenes of an old-fashioned village. If visitors are disturbed by the shreds, they can look up to where the wall meets the ceiling.

Sometimes the destruction is caused by us (although it's a cat's fault), especially when a cat brings a chipmunk indoors and lets her go so he can chase her. Then the race is on, the chipmunk in the lead, the cats right behind her, and the people right behind the cats, all running as fast as we can while our remaining possessions come crashing down around us. The cats are motivated by excitement, the people and the chipmunk are motivated by fear, and it all ends when we manage to throw a towel over the chipmunk, pick her up, and return her to her home in our stone wall.

Life with cats has good and bad sides. If they ruin our wallpaper, furniture, and curtains, they also limit the mouse population, and they do this just for fun—they don't eat most of their victims. Every time I visit the basement I see at least one mouse cadaver with the head bitten off, and sometimes I find three or four. Of course, I pick up the corpses but not every day, so our basement has been called a mouse mausoleum. But mouse hunting by cats is good—each time a mouse is killed, the chance of a mouse gnawing an electric wire and setting fire to the house is reduced by one, so we're grateful.

But the most important reward for life with cats is purring. No sound in the world is as soothing or pleasant as purring. The cats sit on our laps while purring, and as we relax our minds fill with pleasant, dreamlike thoughts, perhaps of wonderful times we've had or beautiful things we've seen. We stroke their heads and behind their ears so they will keep on purring. We so enjoy the sound.

Oh, and as for destruction, as cats grow older they don't cause as much, and the scratch pads you can buy in pet stores help a little. So does the big cat tower I have in my house. It's covered with heavy fabric that takes about ten years to scratch off. We saved an upholstered chair by pinning towels around it, and we've all lived happily ever since.

Cat Tracker

— *Sy* —

"*T*he cat is within fifty yards," my companion, Vermont wildlife biologist Forrest "Frosty" Hammond, announced as he slowly swung what looked like a rooftop TV antenna in an arc, listening for a chirping sound. We were standing in a snowy driveway in Springfield, Vermont. A few weeks earlier Frosty had been tracking radio-collared black bears; I had been in India studying tigers. But the cat we were tracking together lived in the house at the end of the driveway: a seven-year-old, fourteen-pound, black puffball named Darryl.

Though tigers and bears still keep many secrets, the enigma of the suburban house cat remains more mysterious yet. In the 1950s one of the world's top experts on felines, German ethologist Paul Leyhausen, had tried to study the outdoor travels of *Felis catus*. To follow just one house cat "would have required three well-trained, physically fit, and inexhaustible observers, plus a lot more equipment than we could command at the time," he later admitted.

Frosty and I hoped for greater success. We had better equipment: We had obtained, from a company that normally supplied telemetry for studies of wolverines and cougars, a custom radio collar small enough for a house cat. We had a 400,000-candle-power searchlight. And we had Barbara Burns, a Vermont state forester, who had generously provided us with a detailed aerial map of the area we might cover, as well as our study subject. Darryl was one of her two cats, and by her account, spent most of his day asleep. But at night?

We imagined hours of adventure. One English study showed female cats roamed up to 17 acres and males ranged over up to 148 acres. Because Darryl was neutered, it was unlikely he'd be seeking mates. But he could be fighting rivals, hunting prey, or being hunted himself.

Nearly two decades after our Vermont evening, the mystery of cats' outdoor travels remains. Our findings were not exactly revelatory. We encountered several problems—the most salient of which was, every time we went outside to track him, Darryl came to ask us to let him back in.

But now North Carolina researchers have launched the most ambitious effort yet to reveal the secrets of cats' outdoor excursions. Using tiny satellite tracking harnesses, the Cat Tracker project has enrolled a virtual army of cats in a program outfitting them with GPS devices.

The findings will be intriguing and important. Cats affect other species in ways we are only beginning to understand. A controversial 2012 Smithsonian Conservation Biology Institute analysis of previous studies contended that nationwide, outdoor cats kill between 1.4 billion and 4 billion birds and more than 20.7 billion small mammals yearly—making outdoor cats the single largest human-related threat known facing these mostly native wild animals. (Belling a cat can help, but not always; one belled cat, who was also declawed, learned to bat flying birds out of the air and bite them to death.)

That's one reason that the Humane Society of the United States and the American Veterinary Medical Association recommend we keep our cats inside. Another is that indoor cats live significantly longer—to seventeen years or more—than those allowed to roam. With an average life span of only two to five years, outdoor cats are apt to meet violent deaths. One study found 63 percent of recorded cat fatalities were caused by cars. Other outdoor cats die in the jaws and talons of predators or in combat with rival cats. They are also more likely to contract diseases such as feline AIDS or feline sarcoma virus.

Cat Tracker researchers will be investigating diet and parasites in selected cats, too. But mostly, as the website states, "We want to know how cats decide where to go." The data already amassed, from more than 500 cats, are intriguing. Most cats stay close to home, covering fewer than 12 acres in their tracking period. Only about 5 percent cover more territory—but one roamed over 116 acres. And a number of cats were discovered to be "cheating" on their owners. "Many cats, we found out, spend a lot of time at a secondary house," said Troi Perkins, a zoology and fisheries and wildlife conservation student at North Carolina State University who is responsible for downloading the data. "That's quite interesting!"

"We're hoping to find out more about whether cats are cosmically inclined to [visit] other cats outside," Troi said. "Do they have little cat games, or group buddies? Or maybe they're just solo creatures out there."

As for her own cats, Troi has rescued two and rehomed them—where they both stay indoors.

And Darryl? We did manage to glimpse some secrets that snowy night. Every time we emerged from the house with our telemetry, he tried to get back in. Finally, sometime after 10:30 p.m., we let him. But while we had been waiting inside, he had left fresh tracks in the snow. We followed his paw prints down a ravine, into a hemlock forest, over a snowbank, into a thicket. Here he met another cat, then a third, and the three traveled together for some time.

Sleeping Dogs

— *Liz* —

Many people dictate their dog's sleeping arrangement, putting the dog in a crate or chaining her to a doghouse or a tree, but dogs are happier if they can sleep where they like, which usually has something to do with where their owners sleep.

Dogs are social animals, as are humans, so they experience loss and loneliness as acutely as we do. Loss and loneliness are evolutionary tools that help us keep our groups together and were formed long ago, perhaps in the Pliocene epoch before humans existed. If you're small, as were wolves in relation to larger predators, and as were our ancestors compared with the important predators, and if you find your food on the ground, not in the trees, you do better in a group than you do alone. This is true for many reasons, not the least of which is that all of you together are keeping watch for predators. A predator tends to attack from behind. Many social animals sleep in groups but face in all directions, so if a predator approaches you from behind, someone else who's facing you will see it.

Modern members of the dog and human species have found no reason to change their social instincts. If your dog sleeps on your bed, you may notice that he snuggles up to you for a while but may very well face away before he goes to sleep. Not all dogs do this, but most do. That way, one of you will see a predator approaching. You'll understand the advantage if you imagine yourself alone in a Siberian forest or on the African savanna after dark. Would you rather lie down on the ground and fall asleep alone or would you prefer to be with others like yourself, not all of you asleep at the same time and all of you alert to possible danger?

That's why my dogs sleep with me. If predators approach, we'll know. I've always let my dogs choose their sleeping places, and always they choose to sleep with a family member—either a person or another dog. Mostly, though, they sleep with me, either on the bed or right beside it if the bed becomes too crowded. We all feel safe.

My current dogs are small. Going to bed is their favorite activity because I'm lying down and they can see me face-to-face instead of craning their necks and looking upward. They race ahead of me to the bedroom, looking back to see if I'm coming. When I appear, they're standing on the bed waiting for me, so I must push and shove to get under the covers. As I pull up the covers, the Chihuahua dives under with me, waits for me to turn on my side and settle myself, then curls up tight against the back of my bent knees.

But the bigger one, the pug type, stands on my chest and looks down at my face for a while. Normally he doesn't sleep under the covers. He has more wildness than the Chihuahua, and soon enough he turns to face away and settles himself tightly in the curve made by my bent legs and stomach. Each dog has maximized the amount of body contact between us and feels both protective and protected. Sometimes the pug type gives a little sigh. The room is dark and quiet, and soon we're sleeping. During the night a cat also joins us, jumping quietly up on the bed and finding his place on my pillow. He, too, likes the safety and the companionship, but he's a cat, thus not so social.

On rare occasions the pug type senses something. That's why he won't sleep under the covers. Perhaps the local bear is passing the house or our tenant is coming home late and trying to be quiet, which makes him sound stealthy. I'm usually asleep when this happens and am shocked into awareness when the bed explodes. The covers fly off, both dogs are standing up and barking wildly, and the cat is making a dash for the door. But as for me, the highly evolved, world-ruling human, I'm sitting up in confusion. What happened? Is something here? I get up and look out the window. Nothing is here. I see only moonlight and I hear no rustle, no footsteps, not even wind in the trees. But something was here, of that I'm certain, and if it's gone we must be okay.

The dogs watch quietly as I look and listen. If I think things are okay, they think so, too. Even the cat thinks so and inconspicuously returns to the pillow. I get in the bed and we resettle ourselves, but the others fall asleep before I do. I wait for my adrenaline levels to subside.

I'm sometimes asked if we have a burglar alarm. We do—it barks. The skills we acquired in the Pliocene still serve us well, and that's why it's best to let dogs choose their sleeping places. Both of you, or if there's more than one dog or one person, all of you, are better off together.

Feral Cats and Stray Cats

— *Liz* —

Feral cats are born wild or have been homeless so long that they've become wild. Stray cats have been abandoned recently and are trying to live as best they can. It's hard to find homes for feral cats because, being wild, they distrust people, so they may remain skittish for life. It's easier to find homes for stray cats, because they return to a life they know and soon become friendly.

I see all cats as individuals, not as members of a group that may or may not be causing problems, but that's just one way to look at them—and that perspective certainly isn't shared by all. Grant Sizemore, director of invasive species programs for the American Bird Conservancy in Washington, DC, points out that cats are an invasive species. They were moved around the world by people, but our species has no commitment to containing them. They are programmed for hunting whether they need the food or not, and in the United States alone, he says, they are responsible for killing 2.4 billion birds a year. Bird populations are declining for several reasons, and cats are making it worse. Cats also transmit diseases such as rabies and toxoplasmosis, both to people and to other animals.

These are serious matters, surely. What gets me is the accusation that cats belong to the detested group known as invasive species. Humans are also an invasive species that began in Africa but then spread all over the world and are doing vastly more damage to it than the cats. But people do good as well as harm—and so do cats, who are useful in containing rodent populations, which are pests that can also spread diseases to humans. For a while the recycling center in our community encouraged a small group of feral or stray cats that significantly helped with the rodent population. Has a tally been made of how many rats and mice are killed by feral cats in the United States? Probably not.

As I see it, we humans have four possible methods of dealing with feral cats. We could (1) do nothing, (2) find homes for them, (3) kill all of them, or (4) support organizations that help them. As for method 1, if we do nothing, nothing will change. As for method 2 (finding homes for them), many shelters are already filled with cats who aren't being adopted, and these shelters can't take more. As for method 3 (exterminating them), we could try, but there'd be violent protests from cat lovers such as myself. And anyway, such slaughter might reduce an area's feral population for a while, but there's no way the human residents could kill them all, so soon enough the population would recover. As for method 4 (supporting organizations), this would certainly help the cats but would probably not reduce in any meaningful manner the 2.4 billion birds that cats kill annually. As has been said, cats hunt for fun, and it isn't just feral cats who do it. Cats with good homes who go outdoors kill just as many.

An essay of this kind is supposed to end with some sunny resolution that the writer is promoting. I don't have one. I adopt stray and feral cats, I donate to the helpful organizations such as Alley Cat Allies, and I feed any cats who come near my house if they seem to be homeless, but that's about it for me. The sunny part here, if any, is that many helpful organizations do a wonderful job by feeding and neutering feral cats, and they deserve our help and support.

Death of a Dog

— *Liz* —

One of the most devastating experiences we can have is the death of a beloved dog, whose loyalty we never questioned, who loved us all her life, protecting us, helping us, as close to us as any family member, sometimes closer. A dog lives for about fifteen years, whereas her owner may live eighty or ninety years and experience this terrible loss multiple times. It gets no easier.

Some of us then get another dog. The new dog does not replace the dog who died. Nothing could do that. But as is true with friends and family, our circle of love expands without limit. The new dog is wonderful, too. We're charmed, and quickly we become devoted. This is normal. And then, after much too short a time, we lose that dog as well.

As before, we struggle to move on, just as we would if the loved one was a person. But for a person, there's an obituary in the paper, also visiting hours at the funeral home followed by the funeral or memorial service and the burial in sanctified ground. Our grief is understood and honored, gifts to charitable organizations are offered in our loved one's name, letters of sympathy fill our mailbox, flowers arrive at our home, more flowers are placed on the grave, and sometimes a monument is erected or something important is named for the deceased.

But if the dog dies? Nothing. Our mourning isn't acknowledged, we don't get a few days off from work, no flowers are involved, and there is no funeral. The burial is performed by us personally, perhaps alone with only a shovel and our tears. Others will sympathize, of course, but it's we who concern them. They don't mourn for the dog.

So a dog is like a body part, as important to us as our arms or legs. If we're in the bathroom taking a shower and our dog comes in to be together, we are no more embarrassed by her presence than we are by the presence of our legs. But if a person comes in, we might grab a towel, the intruder would back out quickly, and a torrent of apologies would follow—the intruder should have knocked, the door should have been latched, on and on.

Thus losing a dog is like losing a leg. This would change our lives, and our family and friends would be deeply sympathetic, but we alone would miss the leg itself, and our supporters wouldn't know or care what happened to it. We all know where people are buried or what happened to their ashes, but how many of us know what happened to the bodies of other people's dogs?

After the loss, we're very much alone. Most of us don't talk about it. Instead, we keep the dog in our hearts, thinking of her when passing the places where we walked together, missing her warmth when she slept beside us, looking at her bowl, now dry and empty on the kitchen floor. Thomas Hardy wrote a poem after the death of his dog, Wessex, which begins:

Do you think of me at all, Wistful Ones?
Do you think of me at all, as if nigh?

The poem doesn't end well—the dog speaks the last lines:

Death of a Dog

"Should you call as when I knew you,
I shall not listen to you,
Shall not come."

The dog won't come because he can't, but the answer to his question is an emphatic *yes*. Do we think of you at all? We don't stop. We say your name when we're alone, as if you could hear us. We remember the first time we saw you and the last time, too—that moment when we knew you were gone.

The afterlife becomes a question. Some people say that animals go to heaven, while others say they don't. If I arrived in an afterworld and saw only people, I'd know for a fact that my sins had caught up with me, because no place like that could be heaven and only humans go to hell. But I'm not sure there is an afterlife as we imagine it. So I keep the ashes of my dogs with instructions to mix them with mine and put us in the woods. We will then be together until the end of time, at least in molecular form. That's not much, to be sure, but it's something.

PART FOUR

·········•·········

Wild Animals

*W*e're so far removed from the natural world we forget that until the Neolithic period, our lives were like those of all animals. We lived from the land as wild animals live now, so it's useful to consider the realities of that experience. For years we've been told that animals lacked consciousness, didn't think as we do, and didn't have memories. They were nothing like us, we imagined, so that the pronoun for a wild animal was *it*, never *he* or *she*, unless she was a pet.

In this we were guilty of misjudgment now identified as "anthropodenial." That word was created by the famous primatologist Frans de Waal and is the opposite of anthropomorphism, or describing animals as if they had human characteristics. *Anthropodenial* means describing animals as if they did *not* have human characteristics. Pet owners have always known that anthropodenial was a big mistake, and now many scientists seem in agreement.

The observations that follow show the depth of animal thought and, to my mind, establish forever the fact that other animals not only think as we do but also ask some of the same questions. A similar observation is discussed in a previous essay about the lioness who yawned empathetically, showing that as she watched me, she felt similarity.

Other observations, also about lions, are somewhat more dramatic and appear here thanks to the research work of Katy Payne at the Bioacoustics Research Program at Cornell University. As a young woman without formal scientific credentials, she made one of the most important biological findings of the twentieth century when she discovered that elephants make infrasound. Before her discovery, it was held that no land animal other than a certain grouse made sounds too low for humans to hear.

It was my great privilege to be with Katy when she made this discovery and to be with her in Namibia's Etosha National Park as her research continued. One evening we were up on a high platform she had built near a water source, waiting for elephants to show up, when an aging, black-maned lion came walking slowly from the east. Taking no notice of us, he climbed to the top of a nearby rise of ground and lay down facing west, propped on his elbows.

The sun was setting. He watched it. As it approached the horizon he roared, then roared again and again as the sun went down. He fell silent when it went under. He waited for a moment, still looking at the west, then stood up slowly and walked back the way he came.

When I tell people about this observation, they usually insist that he was roaring at another lion. But there was no other lion. He was roaring at the sun. This was also Katy's interpretation of a later event when she was recording lion voices from the top of a tall, wide mesa. Around the base of the mesa were five or six lionesses, spaced over more than half a mile, all of them roaring as the sun went down. This took place on several successive evenings but not every evening.

What was their purpose? Were they were chasing the sun away? Were they warning it not to come back? They didn't like the sun—it made them hot and prevented them from hunting success-fully, because with its light their prey could see them. But that interpretation may seem too simple, and anyway, when a lioness once chased me to get rid of me, she didn't roar: She ran toward me, then slowed down and turned aside in a satisfied manner when I jumped in a vehicle.

Was roaring at the sun a local custom, or do lions every-where do it? It was probably local. Not all lions do it by any means—not even all Kalahari lions did it. When I was camping in other parts of the Kalahari with lions nearby, I seldom heard roaring at sunset, and if I did, whoever was roaring didn't do it as the Etosha lion did—beginning when the sun was perhaps two fingers above the horizon and ending quite precisely when the top of the sun disappeared.

There may be no obvious answer, except that it was a cultural practice of the Etosha lions, but the interesting behavior wasn't unique to them. I knew two captive wolves, for instance, who every morning stood side by side in an east-facing window and sang a song in two parts as they watched the sun rise. They had to see the sun itself, though. They never sang if the sky was cloudy.

Surely most species understand the importance of the sun, and members of at least two species do something about it. Because they look straight at it while they roar or howl, they seem to be addressing it, or at least sending the world a message about it, making their most important, most communicative calls when the sun is leaving or arriving. We may never know what they're thinking as they howl or roar, but that's the kind of thinking that gave rise to philosophy.

—*Liz*

Bears

— *Liz* —

*B*eing dark-colored and the most massive animals in New England, black bears inspire the human imagination and are credited with behaviors that arise from human fantasy. For instance, they are believed to make a hooting call that some people say they can copy, and when they do, a bear sometimes answers them, these people claim.

But bears don't make a hooting call, and if a human tries to make one and hears an answer, the "answer" is probably from another human with the same belief. Where the idea comes from remains a mystery.

Then, too, mother bears with cubs are thought to be exceptionally dangerous. But according to Benjamin Kilham, the world's foremost expert on black bears, attacks by mother bears with cubs account for "only 3 percent of the fatal attacks on humans in the past 109 years." I'm quoting from his spellbinding book *In the Company of Bears.*

This is not to say that it's safe to mess with a mother bear's cubs—doing so could raise the tally to 4 or 5 percent. In the unlikely event that you meet such a bear while on a hike, she will send her cubs up a tree and you will probably be okay if you don't get excited and try not to run but move away quietly with an agreeable facial expression. (Ben Kilham's book offers detailed instructions on what to do in bear encounters.)

So far, since the year 2000, black bears have attacked fewer than thirty people in all of North America. That involves 600 million people and 600,000 bears. Thus we fear black bears without much reason, except that in recent years, danger has increased due to the human desire to take a cell phone selfie while standing beside a bear, an unwise practice that has prompted internet warnings.

Why discuss this? Because each autumn, bears must gain enough weight to support themselves while hibernating. They know that humans treat plenty of their edible material as garbage, so sometimes bears come around our houses in hopes of finding food. Certain researchers have found that if bears are fed by game wardens in predictable areas, they are unlikely to scavenge around people's homes or campsites. Other researchers tell us that feeding bears encourages them to visit human habitations, which frightens the humans, who then shoot the bears or report them to the authorities who shoot them.

Speaking from personal experience, I've seen that bears do very little harm and will move away from a house if asked. One night a bear started to come in our open kitchen window. Her facial expression was pleasant—ears forward as if in interest, lips relaxed—but our kitchen is small and my disabled husband in his wheelchair couldn't move fast. I thought a bear inside would not be good and made a blast of noise with a boat horn I keep handy. The bear withdrew and left.

On another occasion I noticed that the window on the kitchen door was completely dark, as if someone had put a black blanket over it, while through the other window I saw moonlit grass and trees. This seemed strange. I went to look out the darkened window, pressed my nose against the glass, and saw that if not for the glass, my face would be buried in fur. A bear had come for our bird feeder, which normally I bring in at night but had forgotten, so he was standing on his hind legs, leaning against the kitchen door and eating the seeds. I'm just over five feet tall, and I was looking at his lower rib cage, which means that on his hind legs he might have been eight feet tall.

I cherish the experience of being the width of a windowpane from a bear's furry body, but the experience didn't last. When I realized what I was seeing, I drew back quickly, and when the bear realized that he was all but touching a human, he dropped to all fours, tearing down the bird feeder along with the steel pole that held it, which bent like a paper clip. He walked away but came back a little later to eat the seeds he had spilled while I watched him through the window. When I wrote about this incident in a newspaper column, I expected emails from readers telling me what I should have done—maybe called the police or fired a gun—but actually I did what I should have done, which was nothing. I enjoy the presence of bears.

Bears were here long before people. For a while they were eliminated from New Hampshire but they've repopulated, and they're doing well now with no help from us, or not much. May we forever respect them and enjoy their presence, and if we can't do that, may we at least let them live in peace.

The "Dog" We Love to Hate

— Sy —

*T*he fluffy, spotted babies wrestle and lunge, spin and pounce. They're still a little wobbly on their paws, and their furry baby tails still seem as flexible as rubber. The scene reminds me of the first time I met my border collie puppy, when he was six weeks old: He was playing just like this with his siblings at the farm where he was born in a neighboring New Hampshire town.

But I am half a world away, in Kenya, watching through the windscreen of a Land Cruiser, and instead of darling puppies, the animals I am watching are almost universally despised across human cultures. In the company of the world's top expert on the species, Kay Holekamp, I am charmed and captivated by the action at a den of spotted hyenas.

"How can you *not* like hyenas when you see *this*?" asks the Michigan State University biologist. "They're even cuter than puppies or kittens!"

With long, dark muzzles, blond coats with black spots, bristly black tails, and ears that look like a cross between a teddy bear's and an elf's, spotted hyenas might be mistaken for an exotic breed of dog—and not that long ago, some researchers thought hyenas might indeed be dogs' ancestors. Actually they are more closely related to cats. Yet even though they remind us of our most beloved household pets, hyenas get about the worst press of any mammal known to humankind. Hyenas are almost anti-pets: They are characterized as the very opposite of the animals we pamper in our homes.

Ernest Hemingway called the hyena "devourer of the dead, trailer of calving cows, ham-stringer, potential biter-off of your face at night while you slept, sad yowler, camp-follower, stinking, foul. . . ." In 1859 one famous naturalist noted that while villains like Aaron Burr and Judas Iscariot had their defenders, no one would say a kind word about hyenas. Even Disney seems to hate them. In *The Lion King* they're portrayed as cowardly, stinking thieves.

For twenty-seven years Holekamp's observations of hyena families in Kenya's famous Masai Mara National Reserve have been dismantling these stereotypes. Turns out that spotted hyenas are not mainly scavengers; they kill 60 to 95 percent of the food they eat. A single 130-pound female can kill, unaided, a 500-pound bull wildebeest; in fact hyenas—not lions—are, she says, "the most formidable predators in Africa." Lions steal kills from hyenas more often than the other way around.

Though most closely related to mongooses and meerkats, hyenas are more like monkeys in their social complexity, Holekamp has discovered. Spotted hyena society comprises clans that can number more than a hundred individuals. All members of the clan know each other, and each has an assigned rank, inherited from birth. Clan members cooperate to raise cubs in communal dens. They defend a common territory from rival clans. In this way hyena society is sort of like a feudal kingdom, but with a twist: This society is dominated entirely by females.

And even more unusual, the females look like males. It's so confusing that years ago, when a collector was sent to capture hyenas for a zoo, he reported he could find only males ... until one of his captured "males" gave birth in front of his eyes—through a tubelike organ that looked exactly like male equipment.

So why have hyenas evoked horror instead of awe and amazement in most human societies?

Well, they do sometimes scavenge and dig up bones—including those of human corpses. Like most predators they sometimes attack children. And they like to roll in smelly substances—but so do our beloved dogs.

Why some animals are loved—our dogs and cats, for instance—and others hated, eludes even animal behaviorist and *Psychology Today* blogger Hal Herzog. The author of *Some We Love, Some We Hate, Some We Eat* has studied the people-animal connection for as long as Holekamp has studied hyenas, and even he doesn't know. "What you see are these big themes in human nature: part biology, part culture, part voodoo magic that we don't understand," says Herzog.

Ironically, part of the human dislike of hyenas may stem from the very reason they are so fascinating. "Hyenas are just weird," says Holekamp. That may make people uncomfortable. But it's what has kept her studying these families for so long—one of the longest studies of any wild animal in the world.

Hyenas "appear to violate the rules of mammalian biology," Holekamp tells me. "Studying the oddballs can teach you about the basics," she explains. "They allow us to gain insight into what the rules actually are." And by showing us an alternative way to sociality and intelligence, they help us better understand our own beloved pets, and perhaps even ourselves.

Great White Sharks

— Sy —

A few summers ago, on every one of my trips to Cape Cod in Massachusetts, I worried about sharks.

I was afraid I wouldn't get to see one.

That summer I was lucky enough to accompany the state's Division of Marine Fisheries shark biologist Greg Skomal to document his study of some of the Cape's latest newcomers: great white sharks.

But at first the sharks were not cooperating.

Drawn by a rebounding population of gray seals who come to these shores to birth their pups, these most feared and powerful of all sharks show up at the Cape each summer—right about the time human beachgoers get there. What could go wrong?

To hear some people tell it, big, scary great whites are lurking everywhere, waiting to snatch swimmers. "Nothing could be further from the truth," says Skomal. "They don't want to eat people," he explains. "They want to eat seals." On the rare occasions when they bite someone, it's by mistake—and they spit the person out. This is what happened to a Truro tourist in 2012, who was swimming far from shore and close to seals—the first shark attack in Massachusetts waters in seventy-six years (and one from which the victim recovered completely).

"Great whites are not at all what people say about them," Skomal told me. "They're not all curmudgeonly and angry and wanting to kill something. I've never met one like that." In fact, great whites can be surprisingly shy—as I discovered on my first shark sortie with Skomal and his team.

To find the sharks in the Cape's pea-green waters, Skomal and his fellow researchers perform a delicate ballet on sea and air. Flying at 1,200 feet in his single-engine Citabria, pilot Wayne Davis searches for the torpedo shape of a great white in the water. When he finds one, he directs Skomal's boat to film it.

The purpose of the study is to identify as many individual great white sharks as possible, in an effort to find out how many there are. Along the shark's sides, near the gills, pelvic fins, and base of the tail, the meeting of the shark's steely gray upper surface and its white underbelly forms a distinctive, individual pattern.

Filming this sounds like a tall order—and by the end of our second, unsuccessful sortie together, searching sharkless, bucking waves cloaked in glaring sun, I was worried we'd never get the chance to try.

But then came our third sortie, in August. Everything changed: Davis spotted shark after shark. Again and again our captain piloted the boat directly alongside one huge, silvery shadow after another. Skomal's delight was infectious. "Sweet!" he cried, recognizing one of the sharks he'd already tagged—a male named Chex. "Big shark—big boy!" he shouted, as he filmed an exceptionally impressive fourteen-footer. The team captured GoPro footage of six different sharks that day—five males and one female.

The thrill of that day made me long for even closer contact. Later that fall I got my chance. I found myself in an underwater cage, breathing through a hookah, just off the coast of Guadalupe Island, Mexico. Its clear blue waters afford exceptional views of this storied predator.

Inside the cage, I understood I'd be perfectly safe. But what would it feel like to be just yards away from a 1,500-pound fish whose three hundred serrated teeth were capable of severing the head of a twenty-foot bull elephant seal in a single bite?

As I donned my scuba gear to descend into the cage with my fellow divers, my heart pounded.

For the first few minutes, small striped fish called fusiliers swam into the cage; bits of tuna bait floated by; a tiny jellyfish stung my cheek. We swiveled our necks, willing a shark to materialize.

And then, literally out of the blue, from about a hundred feet away, the ocean seemed to gather itself into the shape of a shark, and it swam toward us. Sleek and sinuous, silver above and cream below, the shark was as elegant as a knight in white satin. His dark eye swiveled in its socket to glance at me, then flicked away. There was no menace in his glance.

In that moment I shared Skomal's devotion to a fish many people wrongly fear. Earlier, on one of our sorties, he had described the typical demeanor of these massive, powerful fish: "They're laid back. They're calm. They're beautiful. I want these sharks to survive."

Mysterious and misunderstood as apex predators, for millions of years great whites controlled the balance of the ocean ecosystem. On humans' watch, we have decimated shark populations: We kill 100 million yearly. By 2050 we will have filled the sea with more plastic than fish. No wonder, then, that when that great white approached me in the shark cage, instead of fear, a great sense of calm swept over me. With him in charge, the ocean would be in good hands.

Feeding Deer

— *Liz* —

Most authorities on deer, especially those in the New England fish and game departments, strongly advise people not to feed deer in winter because this can kill them. However, far fewer deer are killed by being fed than by being shot—in 2014, the time of this writing, hunters "harvested" more than 19,000 deer just in Massachusetts and New Hampshire, not counting deer who were wounded but escaped the hunter to die later. But the same authorities point out that at least in New England, deer have few natural predators and would overpopulate if hunters didn't do their part. Then, too, in severe winters deer die of starvation—in spring you find circles of their hair after other animals have eaten their corpses. Even so, there are good reasons not to feed them.

I have great respect for the fish and game departments, and I'm grateful for a wealth of material from David Stainbrook, the deer and moose project leader from the Massachusetts Division of Fisheries and Wildlife, who generously sent me links to various publications showing the damage we can cause by feeding wildlife. According to these publications, winter feeding causes deer to congregate, which may spread diseases, and also to run across roads where they are hit by cars. But most important, deer adjust their digestive systems for winter and are set for diets high in fiber but low in carbohydrates, so the sudden introduction of large amounts of foods low in fiber and high in carbohydrates, such as corn and other grains, seriously disables their digestion and can kill them. In New Hampshire, for instance, the rumens of a group of dead deer were stuffed with apples, corn, and hay. These animals probably died within hours, although others might live for a month or so while some might be permanently compromised. There is no treatment.

For years I've fed deer with some success, but if potential deer feeders knew what it took to do this, the difficulties alone might discourage them, never mind potential consequences to the deer.

For one thing, the timing is tricky. One must start feeding before the deer's digestive systems are completely in winter mode but after the end of hunting season because attracting them to food would expose them, and these dates don't always coincide. One also must recognize the deer as individuals so that one can know if new deer are joining, and must monitor their droppings for diarrhea or other abnormalities and also clean up the droppings to lessen the risk of disease. One cannot just leave a large pile of food and let it go at that—one must know how many deer are eating and put out a corresponding number of moderate piles, always on clean snow, often twice a day, which means that from November to April one can't leave home from for more than a few hours. And one must be prepared for heavy expense that isn't tax deductible. My deer feeding costs about $1,000 a year.

But the worst part is the uncertainty. In 2013 I fed fifteen deer until April 2. The snow was melting and grass had appeared, and the usual deer were in our field grazing. Then I noticed five more deer, a new group, whose digestive systems were almost certainly in full winter mode and could be harmed by a new diet.

They had come for the grass. The year I planted oats, fifty-five deer showed up together on the day the oats sprouted and not an oat remained—but the corn I'd put out was already in place. At least no deer would get much, and I couldn't keep watch all night, but the worry was obsessive just the same. The next day deer were eating grass where snow had newly melted, and none came for the corn, although three of them thought about it.

My reward for this—staying home for five months; dragging heavy buckets over icy, rough terrain in any weather; scraping up deer droppings; and impoverishing myself—comes in the spring when I see from a window the same deer I saw in the fall. Few would find this sufficiently rewarding, especially if the winter had been moderate, in which case they'd see those deer anyway. I've been asked by readers and by those involved with fish and game departments to discourage people from feeding deer, and if this doesn't do it, I'd say that except for the most devoted wildlife helpers, it can't be done.

Which brings us to the question of why help wildlife anyway? I do it because we're so happy to damage them. We destroy their ecosystems to build houses, we hunt them just for pleasure, and we kill them with our cars when they try to cross the roads. If during a harsh winter I can help a little in a responsible manner, whereby the same deer who came in the fall are there in the spring to hide their fawns in the grass on my field, I feel I've done some good.

The Lion

— Liz —

We know less about the minds of animals than we know about their habits, and the better we understand this, the deeper their mysteries seem. The truth of this was demonstrated by a lion and Katy Payne, a researcher at the Bioacoustics Research Program at Cornell University who, as mentioned in an earlier essay, discovered that elephants make infrasound.

Having spent most of her life studying wildlife, especially whales and elephants, Katy is more than familiar with the natural world. Her experience described here took place when she was in Namibia, hoping to record the roars of a lioness who had been on the far side of a fence that circled the tourist area of Etosha National Park. The lioness wasn't in view at the time, but she'd been roaring, and Katy thought she'd roar again during the night. So with a recorder in hand, Katy was in a sleeping bag inside the fence, lying on her stomach, waiting.

But instead of the lioness, an enormous lion with a yellow mane walked up to the fence. Katy had noticed a hole in the fence, and so had the lion. The hole was patched, but only with chicken wire, and he sat down right in front of it.

Katy describes this adventure in her fascinating book, *Silent Thunder: In the Presence of Elephants*, from which I'll quote to recount the tale. Her account, however astonishing, is true. The lion wanted to look at her. He may not have known what the sleeping bag was, or he may have seen that a human was in it and wanted to know if she was easy prey. As he sat down he stared at her face, perhaps wondering what to do next. Stretching her back to look up at him, she propped herself on her elbows and met his eyes.

If the lion chose to reach her, he could. She knew if she moved he might scoop her through the hole in the fence with his huge, powerful paw. So she stayed still and kept looking at his eyes, although soon enough her back and arms were getting pins and needles.

The lion also sat quietly. He was "relaxed and alert," she says. The moon was up and creeping higher, "lighting the hair on his shoulders, his whiskers, and the long hairs of the mane around his face."

After a while she saw him panting mildly as a long stream of drool slid off his tongue. On moonlit nights lions often go hungry because their intended prey can see them. And there was Katy, lying still, within easy reach, a satisfying meal if he didn't have to share it. Why didn't he reach through the wire and grab her?

Katy says he had acknowledged that she wasn't food. But she certainly made him think of eating, so was he figuring out how best to grab her and drag her through the fence? It would seem, for whatever reason, that he didn't want to, or not then.

The night passed slowly. Katy didn't move. Instead, she watched the lion's eyes as the moon crept toward the zenith. At first its light shone in her face, but as it crossed the sky its light came from behind her and shone on the lion. Then she could see him in detail. "His eyes," she says "were exquisite. Brown and gold. Neither of us blinked."

It's hard to imagine holding a painfully uncomfortable position for six or seven hours, but by the time the moon reflected in the lion's eyes, Katy had done this. She never shifted her position and her gaze never faltered, nor did the lion's. Etosha Park abounds with other animals, and some must have been rustling in the bushes as the hours went by, but even the rustles didn't distract the lion. He kept looking into Katy's eyes while she looked into his.

As the moon went down, the sky grew light behind Katy and dark behind the lion. Her shadow fell on him. And as the moon sank lower, the sky grew darker. The ache in her arms and her back must have seemed intolerable—by then she hadn't moved for almost twelve hours—but still she didn't take her eyes from the lion's.

As the moon was sinking the sky turned slightly gray. Morning was coming. The lion stood up, stretched himself thoroughly, and yawned. Giving Katie a parting glance, he turned his back and walked off in the bushes. Katie stood up, too, and as she was leaving she heard a terrible scream. She looked through the fence. The lion had killed a kudu.

Etosha lions were known for killing people. During the struggle for Namibian independence, their man-eating habits were blamed for keeping the revolutionary soldiers stationed in Angola from entering Namibia from the north. So if man eating was acceptable to that lion population, why didn't this lion eat Katy?

While contemplating her for those long, uninterrupted hours, he was learning important things about her. Perhaps she seemed balanced like himself, not panicked like a kudu, but what does this mean? That someone must act like prey to become prey? Unless you're angry, and he wasn't, you don't just casually kill those you know and eat them—you interact with them in an acceptable manner.

Surely Katy's demeanor seemed important. She must have seemed unafraid, and like the lion in an earlier essay who roared at the setting sun, she was lying on her stomach, propped on her elbows, watching with concentration just as the sun-watching lion had done. Was Katy doing what a lion might do when watching something? So did the lion watching her think she was like him?

There's another possibility. A number of studies, including studies done by the British Psychological Society, show that looking into another person's eyes for just a few minutes can have important consequences, from discomfort and hallucinations to falling in love. In any case, it always has consequences, usually positive ones, so it's recommended as therapy for couples with marital problems and also is used for meditation.

In most of these studies the participants sat close together, and in some their foreheads were touching. This was not the case with Katy and the lion, who were at least ten feet apart. And it's hard to imagine falling in love with a hungry, drooling lion who is staring at you.

But what about the lion? A study by Japanese scientists showed that our brains synchronize during mutual staring. After a few minutes both people begin to blink at the same time, as if controlled by the same brain. So while staring for hours into Katy's eyes, did the lion experience a sense of oneness? It seemed to Katy that he might have. "We were both in the same place," she said.

The sun changed this, and no wonder. For those who live in the natural world, the life-changing times are sunset and sunrise. The lion went off to kill a kudu and Katy went back to her camp with her story. Both resumed their normal lives and that was that.

I heard a similar account of a woman who was backpacking alone in cougar country when a cougar jumped on her from behind. She fell face down but managed to turn over with her backpack between her and the cougar, who lay on top of her. She looked into his eyes as Katy had looked into the lion's, and may also have spoken to him in quiet tones. At any rate, after a long time while he met her eyes, he, too, might have felt respect and perhaps connection. At last he stood up and left.

If I had been in Katy's situation, I'd be dried-up lion scat by now because the last thing I'd want to look at would be a lion's gleaming eyes. So perhaps the best way to view an animal is to view it as a person in a different form. Our species is just one among 8.7

million others. How many of these can we name? How many do we know or understand? It could be hard to find an animal who would let us stare into his eyes, but if we could (but hopefully not an apex predator), we'd learn more about the minds of those with whom we share our world.

Hyraxes

— Liz —

Sy and I recently went to Tanzania for a wildlife safari. We traveled across the Serengeti to see the wildebeest migration, which is one of the wonders of the world. We were with Richard D. Estes, the foremost wildebeest scientist, and we saw some amazing sights, not the least of which was thousands of herds of wildebeests, maybe fifty to a hundred in every herd, all traveling together, heads low, tails flicking, as far as the eye could see. They were slowly walking to Kenya where they hoped to find green grass.

Our trip had a scientific purpose, but we didn't deny ourselves a little wildlife tourism, whereby you ride around in a vehicle looking out the window, stopping when an animal appears, soon moving on to see another animal. Thus it's somewhat like sitting on a sofa, turning the pages of a coffee-table book. On the first page you see an elephant, on the next a vulture, on the third a giraffe—on and on, day after day. Such safaris are for viewing animals, not for understanding them, and only if someone like Dick Estes is with you can you do both. We saw thousands of animals. We learned from Dick how wildebeests transformed much of Africa, creating and preserving the open savannas by traveling over them and grazing, and we saw for ourselves how this was being accomplished.

I was fascinated with wildebeests. I also became fascinated with hyraxes (*hyrax* is pronounced HIGH-rax), but I hadn't known much about them until the vehicle we were using broke down and needed repair. That day we stayed at a safari camp in a somewhat forested area that happened to be hyrax habitat, so I watched these animals all day.

What is a hyrax? It's a primitive little mammal sometimes called a coney, found across much of Africa and the Middle East. The ones I saw were rock hyraxes, about eighteen inches long with short legs, short tails, and round, plump bodies covered with gray fur. They had tusks like elephants, but the tusks were small and hardly showed. Otherwise their faces were something like a squirrel's. They ate leaves and grass, which they bit from the stems by twisting their heads to use their molars. And because they were primitive mammals, their thermal regulation was underdeveloped, so despite their thick fur, they needed the sun.

The camp where I watched them was on high ground (maybe eight thousand feet), so the nights were cold and little groups of them had cuddled together in sheltered places. When the sun was up, they spread out to open places where they lay on their sides and exposed their white bellies to the sun's warmth.

Hyraxes are mentioned in the Bible (Leviticus 11), where God instructs Moses and Aaron to tell the Israelites that "the hyrax though it chews the cud does not have a divided hoof: it is unclean for you." Hyraxes don't chew the cud, so one wonders why God thought they did, but his view of their paws was accurate, as these are something like our feet—flat and relatively long, with toes and a heel. Hyraxes are said to be the closest living relatives of elephants and are equally intelligent or nearly so.

Soon I was so taken with hyraxes that I longed to have one as a pet. I loved the careful attention they paid to their surroundings, interested but unexcited. I loved the way they ate, chewing slowly and thoughtfully. I loved the way they related to one another, casually aware but not particularly engaged, as if they knew and trusted one another. Elephants act this way, too, if no one is bothering them, and who wouldn't want a small, furry creature who behaved like an elephant as a pet? I would never catch a wild animal to keep as a pet, of course—removing him from friends and family and the ecosystem he understood so he could spend the rest of his life in a cage—so I will content myself by relishing the knowledge that hyraxes once ruled the world.

This was in the Eocene epoch, the age of hyraxes. As the dominant group, they spread through Africa, the Near East, Asia, and most of Europe. Their numbers included carnivores and herbivores—some enormous, some midsize, and some small. Some became aquatic in the manner of beavers, and some of these gave rise to elephants and manatees, or so it's said. And even today, except for their size and appearance, modern hyraxes can remind one of elephants, tusks and all, in the way they behave when at peace by themselves.

As for our safari, despite the lions, the wildebeests, the crocodiles, and the vultures, if hyraxes were all I'd seen, just to have them as a memory would have made the trip worthwhile.

Christmas Ermine

— Sy —

Every Christmas, for the holiday, I bring our flock of hens a brimming bowl of hot popped corn for breakfast. It is greeted with great enthusiasm. But one Christmas morning was different. I opened the door to their coop and found one hen lying dead on the wood shavings carpeting the floor. Everyone was subdued.

Some of our hens were elderly at the time. It was possible, I thought, that she had just keeled over from her perch from old age. I reached down to pick her up by the legs to examine her. Her head was wedged into a small hole in the corner. But I couldn't lift her. Something—or someone—had a hold of her head.

I pulled and pulled, and finally yanked my chicken free. The next second, out from the hole popped a white face less than an inch wide with a bright pink nose and coal-black eyes burning with intensity. It stared directly into my eyes. It was an ermine. I had never seen one before. Instantly my sorrow was replaced with wonder.

The tiny animal before me was gorgeous. Its fur was the purest white I had ever seen, whiter than snow or cloud or sea foam—so white it seemed to glow, like the raiment of an angel. It's easy to see why kings (and even Saint Nicholas) trimmed their robes in ermine fur. But even more impressive was its gaze, a look so bold and fearless that it took my breath away. Here was a creature the length of my hand, who weighed little more than a handful of coins, but who had come out of its hole expressly to challenge a monster who was a thousand times its size. *What are you doing with my chicken?* those coal eyes said to me. *Give it back!*

Of course, I had been thinking it was *my* chicken. I had raised her from an egg-shaped chick from the time she was two days old. Our chicks grow up in my home office. They snuggle in my sweater and perch on my shoulders as I write; when they grow older, they follow me around outside. They come running to me when I call. I love each one, and I loved the hen who lay dead in my arms. Her body was still warm. But I could feel no animosity as I faced the one who had killed her. I was gobsmacked.

"There is something enormously satisfactory about a weasel," New Zealand researcher Carolyn King writes in her book *The Natural History of Weasels and Stoats*. "It has the perfection, grace and efficacy of well-designed tool in the hands of an expert." New England has several species of weasel, all just a few inches long, the smallest of which (the least weasel) is only as long as a man's finger. All are brown with light bellies in summer, and when they turn white in the winter, they are known as ermines.

These are the world's smallest carnivores. It is as if all the ferocity of the world's wild hunters—lions, tigers, wolverines—has been concentrated and compacted into a creature smaller than a vole. Quick as lightning, an ermine can leap into the air to kill a bird as it takes flight, or follow a lemming down a tunnel. It can swim, climb trees, and bring down an animal many times its size with a single bite to the neck—and then carry it off with a bounding run. An ermine consumes five to ten meals a day. It needs to eat least a quarter to a third of its own weight just to survive in captivity, and much more in the wild, especially during the cold winter. These little animals' hearts beat nearly four hundred times a minute. No wonder they kill everything they can at every opportunity.

The ermine held my gaze for perhaps thirty seconds. Then it popped back into the hole. I desperately wanted my husband, Howard, to see it. What were the chances the tiny animal would still be there, much less show itself again, when I came back? Yet I put down the hen where I had found her, ran the five hundred yards back to the house, alerted Howard, and then together we returned to the coop. Again I picked up the hen. And again the ermine shot its head from the hole, its black eyes blazing from that luminous white face as its piercing, fearless stare met our eyes.

Even in the wake of tragedy, we could not have felt more amazed had we been visited by an angel that Christmas morning. In our barn we had, in fact, beheld a great wonder—as the magi had in the barn that they had visited so long ago. Our Christmas blessing came down not from heaven but up from earth. With its dazzlingly white fur, hammering pulse, and bottomless appetite, the ermine was ablaze with life: so pure and so perfect that in its presence, there was room in our hearts for neither sorrow nor anger, just awe.

One Mouse

— *Liz* —

This is about a mouse who was lying unconscious under our refrigerator. Perhaps a cat had caught her outdoors and brought her in through the open window or found her in the basement and brought her upstairs and then lost interest when she fainted. I thought she was dead and wanted to throw her away, but I couldn't quite reach her, so I went to get a broom. When I came back she had moved to another place, but she still seemed unconscious. So I put her little body in a cage with food, water, and cotton balls to make a nest, although I thought she would die during the night.

But by morning she was better. I put the cage in a closed room with the cage door open. She ran out of the cage and hid under the radiator. That night she came out to explore the room. I'd intended to leave the door open so she could depart if she liked, but because the cats might find her, I made another plan. She had an appealing personality, she represented one of the world's most successful species, we shared a common ancestor, and her direct, more recent ancestors had connected our species to cats. Who was I to break with tradition? I brought her more food, water, and cotton balls and shut the door. My plan was to keep her as a pet.

When this account was published as a column, I expected angry mail from readers pointing out that mice are pests who carry human diseases and should be exterminated, even though humans carry more human diseases and we don't exterminate one another, or not for that. "Find a phone," I planned to tell such readers. "Call someone who cares what you think."

Mice have been here longer than we have. We share a common ancestor and much of our DNA. To keep this mouse was almost like helping a relative. Or sort of like helping a relative. One night I stayed a long time in her room after bringing her food, standing still, hoping to see her. When a tiny ant tried to cross the floor, she ran out from under the radiator, ate it, and ran back. This little mouse might have enjoyed the help I gave by feeding her, but she certainly could feed herself.

I seldom saw her in the daytime or when the lights were on at night. Sometimes by moonlight she would run along the wall, glimpsing me as she went speeding by. But sometimes she'd stand still so we could consider each other. Our size difference seemed important—she was three inches long, I was sixty-three inches long; she weighed less than an ounce, I weighed 140 pounds; she was the size of a walnut, I was the size of a leopard—but even so, we watched each other. We both had brown eyes.

Sometimes I'd leave the door open, having confined the cats, but the mouse knew they were out there and stayed in her room. Mice live in colonies, and our house may have had two. I didn't know which colony she belonged to, but if she found herself in the wrong one, its members would harm her. Maybe that's why she stayed in the room. The radiator was warm, the food was plentiful, and if I came in now and then, perhaps that wasn't so bad.

Even so, as time went by I realized that I liked her more than she liked me. We humans are not the only species that connects with others—we're merely an example of how this is done—and in various ways many animals have connected with me, including rats. But rats have very different personas than mice (there's a reason why calling someone a mouse means one thing and calling that person a rat means something quite different), and this might explain why the mouse and I were not connecting.

Or did our size difference explain it? Even elephants are not as big in proportion to me as I was in proportion to that mouse. Also, elephants stand on all fours so they can lower their heads and put their trunks down where we smaller creatures can touch them. If I were to make a similar contact with the mouse, I'd need to lie on the floor, but even if I put my chin on the floor and she stood on her hind legs facing me, her head would only come up to my nose and my eyes would be way above her.

Three weeks passed. I could see from her droppings that she spent most of her time under the radiator, and we'd made no connection, so I gave up. I got my longest scarf out of the drawer and draped it over the sill of the open window in the mouse's room. One end of the scarf was on the floor and the other end was on the grass which, as it happened, bordered a stone wall and a collection of bushes. She could stay outside if that's where she'd come from, or if she wanted to be indoors she could find one of the many small entrances to our basement that mice use in the fall.

I gently closed the door of the room. When I opened it a little later, the mouse was gone. For several days I left the window open with the scarf in place in case she changed her mind, but even after all I'd done for her, she wanted freedom more than safety and food, and far more than she wanted human company.

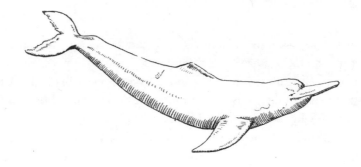

Pink Dolphins

— Sy —

"*I*n Iquitos, many years ago," our Amazon guide, Moises Chavez, told me my first day in Peru, "before the Belen market was built, Indians lived there, on the banks of the river." On Saturday nights the young men and women from the village would gather to dance beside the river. What the people didn't know, Chavez told us, was that a dolphin was watching them. And the dolphin fell in love at first sight with one particular girl.

"Some people say the dolphin's just a dolphin," Chavez said. "But the Indian people, they know a different story."

It was the stories about the strange, pink dolphins of the Amazon River that first drew me to Brazil and Peru. People say the dolphins are magic. They claim the animal can change into human form and shows up at dances as a handsome stranger. The dolphin can seduce you and may take you away to the *Encante*, the enchanted world beneath the water. Those who visit this world seldom return, because everything is more beautiful there.

The stories sound impossible. But so does a pink, river-dwelling whale. Yet these dolphins really can be pink. Not all are; some are grayish or pale. But some are as pink as bubble gum. And like all dolphins, they really are a kind of whale—but this whale lives in fresh, instead of salt, water. I know this is true, because by the end of my four expeditions to the Amazon, I had observed pink dolphins for hundreds of hours.

And I, too, fell in love with them.

Unlike marine dolphins, these creatures don't leap from the water in spectacular displays. Their athleticism comes from their balletic flexibility, which they need to navigate often dark waters that rise, during the wet season, high enough to engulf trees. They must fly like birds through their branches. Their pectoral fins look like wings. Their faces, when they poke through the water to look at you in your canoe, look rather like ours. That is, until they open the tops of their heads and gasp, "CHAAAA!"—and then dive.

How could I not follow?

In fact my first plan, before I arrived in South America, was to follow these dolphins on their migration. I had read preliminary studies by researchers Vera da Silva and her late husband, Robin Best, suggesting these dolphins travel seasonally across international borders. I proposed to follow them to find out where they go. But when I arrived in the Amazon and met da Silva, she told me that the suggestion she made in the paper had been proved wrong. They aren't migratory.

So I followed them in a different way. I followed them not from point A to point B but as a disciple follows a leader. Or as a lover might follow the beloved.

My travels with the dolphins led me back through time. Following their lineage through prehistory, I discovered they are the remnants of an ancient whale lineage that entered the Amazon River from the Pacific—before the rise of the Andes Mountains 15 million years ago. And even further back, thanks to the pink dolphins, I found that the whale lineage grew more surprising yet. The ancestors of whales once walked on land; the first aquatic whales, whose fossils were recently discovered, had small, vestigial hooves.

I followed the dolphins through the stories of the local people. Chavez told me a story he had learned from his grandmother when his family had lived in the remote Napo River region among the Yagua Indians.

"One day—a day like today—they gonna celebrate a really big party," he began. That night a handsome stranger, with light skin and blue eyes, showed up at the dance, and he gave the girl he fancied a big diamond and a gold watch. He asked her to tell no one, and that he'd meet her again at the next dance. She agreed. But the girl broke her promise, and she bragged about her gifts to her father and brothers. At the next dance the men tried to restrain her lover. But he broke free and dove into the water. When he disappeared, her ring turned into a leech; her gold watch turned into a crab and crawled away; and the girl was heartbroken and lonely forever.

I heard dozens of stories like this one all over Peru and Brazil. Many people—including one of da Silva's assistants—told me they had actually met enchanted dolphins. One had nearly been seduced!

So are the stories true? Is the *Encante* real?

It's true that the world beneath the river harbors wonders more spectacular than our wildest dreams. Scientists tell us that in the Amazon basin there may be plants and animals that offer cures to our worst diseases; they insist hundreds of new species await discovery. Many species we already know about surely qualify as magical. Where but in an enchanted world could you find pink whales who fly between the branches of treetops?

But this magic place is under siege. Like the girl who betrayed her pink dolphin lover, we're forgetting our promises to the natural world. Due to timber interests, mining, cattle ranching, and fires, half the Amazon's rain forest could be lost by 2050. Dams built across the Amazon basin block migratory routes for fishes and ruin animal habitats. Pink dolphins are even being slaughtered for catfish bait.

The pink dolphin stories ring true in a very deep sense: It's important to honor the connections we share with our fellow species. If we do not, the world loses its enchantment, and we risk destroying what we love most.

Happy Rats:
Playful, Ticklish, Optimistic

— *Sy* —

"*H*oney, can my friend the anaconda tamer come and stay with us for two days?" I asked my husband.

"Is she bringing an anaconda?" he asked, alarmed.

"No, the anacondas live at the aquarium," I replied. "But she's bringing her six rats."

"Six rats!" he exclaimed.

"That's why she can't stay at the inn," I explained. "And one of the rats is sick. So she can't leave them at home."

In the end my gracious husband welcomed our multispecies guests, but only two of the rats showed up. Then Liz was upset. "What happened to the other four?" she asked. She, like me, had wanted to meet all six and was disappointed that my visitor had left four behind.

Rats bring out vastly different feelings in people. Those who despise rats are usually laboring under one of many rat misconceptions—like the recently exploded myth that rats brought the black death to Europe. (Maligned for eight centuries, black rats were recently exonerated when University of Oslo scientists discovered the culprits bringing plague-infected fleas were really Asian gerbils.)

Those who appreciate rats tend to be folks who can look beyond an animal's reputation and see the creature's true nature—people like my friend, the anaconda tamer, Marion Lepzelter. (And no, she does not feed rats, or any live animals, to the anacondas.)

"Rats are as smart and affectionate as dogs," Marion assured me as she took Zero, a white rat, age two, out of her sturdy travel cage for me to hold. Soft and warm, Zero looked me in the face quizzically, whiskers questing, eyes bright and curious. I offered her some yogurt on the tip of my finger. She licked it speedily and cleanly with her tiny pink tongue, and thereafter we were fast friends. "This is Sy," Marion told her, just as if she were introducing two people.

Rats understand some human words. They easily learn their names and will come when called. They can fetch, walk a tightrope, and sit up, among other tricks. Marion has taught several of her rats to "play basketball," eagerly carrying a ball toward a mounted hoop, standing up, and then pushing the ball through the hoop with their tiny, dexterous hands. A Belgian charity has trained giant African pouched rats to detect landmines and diagnose tuberculosis.

"Zero's just a plain old white lab rat, but she's the snuggliest rat I ever had," Marion told me. As she chucked Zero under the chin, Zero closed her pink eyes in ecstasy.

Zero came along to provide comfort and companionship for Pepper. At age three, Pepper, named for her dark fur, is ancient. She's blind, missing some teeth, and like many elderly people takes a variety of medicines, including Viagra (which was originally developed as a heart drug). But she still relishes loving touch. When she's not cuddling with Marion, Pepper snuggles with Zero, who is her best friend.

All rats—the genus *Rattus* has more than sixty species—love to snuggle. Rats are social animals. They help other rats in trouble, establish friendships, sleep together, groom and play with each other.

And like people, rats laugh when they're happy. Using a bat detector to make their ultrasonic chirps audible, neuroscientist Jaak Panksepp discovered this in 1990. Rats laugh when they play with each other, and they laugh when they're tickled. Laughing rats playfully seek tickling from their people just like dogs urge their people to play.

In later experiments Panksepp discovered that happy, laughing rats are more optimistic. He trained rats to associate pushing a lever that yielded a treat with one tone, and a lever that could avert an electric shock with a different tone. He then tickled one group of rats and merely handled another. Then he played a third tone, one that sounded similar to both the others. The rats who had been tickled rushed to press the lever that yielded yummy food. Because they had laughed and were happy, they expected good things from life.

Perhaps we can learn from them, and from Pepper and Zero. When we approach an animal expecting it to be clean, intelligent, and friendly instead of dirty, stupid, and mean, more often than not we may find our expectations rewarded—and begin to enjoy the company of an animal we may have previously feared.

PART FIVE

———————◆———————

Tiny Animals

I f you consider our species in relation to others, we're midsize,
meaning we're big enough to be seen at a distance but not
as big as a whale. This is also true of the species with which
we're most familiar. For example, when a horse or dog walks by, we
instantly know what we're seeing. But most of the life-forms we live
with are small. We usually don't see them, don't know what they are if
we do see them, know next to nothing about them, and have little if
any interest in them. If one of them walks across the table, someone
usually whacks it, not caring what it was or why it was there.

Such mindless hostility describes our relationship with all tiny
animals except butterflies and bees. At least we know what these are,
sort of, in that butterflies are pretty and bees make honey, so these
little souls have value in our eyes, unlike the thousands of other small
animals surrounding us who are unknown to us, however remarkable.
Anyone's lawn, for instance, could harbor hundreds of tiny creatures.

Do we ever ask how someone so small survives a winter? We
don't, so we're surprised to learn that some of them freeze solid.
Consider the woolly bear, for instance—the larva of the tiger
moth. This little caterpillar hatches from his egg in the fall and
roams around until winter, which freezes him solid. It's said that
you can ring a bell with him. In spring he'll wake up to do his
caterpillar duties, learning then what he needs to remember for the
rest of his life. This can go on for several years until he pupates and
emerges as an educated moth. He and many other tiny creatures
take refuge in leaf litter dropped by the plants in the fall, and some

of the others also freeze solid. This means it's wrong to rake leaves from your lawn in the fall. You should rake them only in the spring, after the tiny inhabitants are up and moving.

Among these small creatures are the single most successful animals that ever lived—the water bears who will be mentioned later. They've been on earth for 500 million years, and their phylum has come unscathed through every known extinction period, however drastic. They'll be here long after the rest of us have gone extinct, and most of us have never heard of them.

Worms are another example. How many of us know about worms? We think we do—we think they're good for the soil and all that—but as it turns out, such knowledge while accurate is minimal. One of the greatest mysteries in biological science involves the simple earthworms I see on my driveway after a rain. They're not native to this country. You'll see what I mean when you read the essay about them.

All in all, if the smallest animals were big, we'd understand them better, we'd be used to their appearance, and we'd admire what they do. These tiny creatures can learn, as has been demonstrated scientifically, and they also have emotions such as fear and even frustration or disappointment, as another essay will explain. I once watched a tiny beetle trying to make a decision, but I was with other people, one of whom brushed it off my sleeve, which it was trying to explore. That ended my observation, or I'd describe it.

Imagine looking at a soaring eagle, high above us in the sky, a noble, magnificent creature. If the eagle were the size of a fly and flew around us, we'd spray it with one of the many lethal chemicals developed for that purpose. These smallest animals will never be big enough for most of us to feel any connection, but the better we know them, the more we can appreciate the paths that evolution has taken to create this enormous complexity of life-forms. Only one of these life-forms is us.

—*Liz*

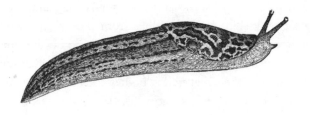

Slugs

— *Liz* —

*T*he names of animals have different connotations. To call someone a pig is fairly insulting, to call someone a louse is highly insulting, and to call someone a slug is exceptionally insulting, implying laziness, stupidity, and uselessness, so I'm wondering how readers will react when I say I'd like to have a slug as a pet. "You want a slug?" someone will cry. "What for?"

Let me explain. After a brief, much-needed rain, I was behind our house checking the rain gauge (we got almost an inch that time) and what did I see but five slugs in the still-damp grass! Two of them were mating—one was on top of another one, anyway—a third slug was at a slight distance, perhaps also thinking of mating, and a fourth and fifth were a few feet away, both minding their own business and evidently relaxed as if they had already mated. Years had passed since I'd seen a slug or even thought about them, so I was excited.

Our planet has many kinds of slugs, which are related to snails, clams, oysters, and octopuses. I don't know which kind mine were except that they're known as gastropod mollusks, which means "stomach-foot mollusks," because like snails they slide along by contracting the skin on their bellies from the waist down while stretching the skin from the waist up, then doing the opposite. Somehow they flow forward smoothly and surprisingly quickly, considering that their featureless bodies look the same from all directions and seem even less complicated, say, than a banana. Their bodies are versatile, however. Like an octopus, a slug can push its body through the tiniest hole, a little at a time. And when they move they squeeze out a small amount of slime, which helps them glide quickly and also protects their undersides from roughness.

But often enough when you look at a slug, you see nothing more than a tiny brownish oval shape that isn't moving. So what surprised me when I saw the slugs in question was their awareness, especially from the two slugs mating, as if the slug on top felt the need to understand his surroundings. Slugs are hermaphrodites, having both male and female organs, but the slug on top seemed to be taking the male role. His rear end (I use *his* because of the role he was playing) seemed to be adorned with tiny, dark protuberances not evident on the slugs that weren't mating, and was partly wrapped around the slug below. He seemed to feel responsible for the safety of them both because he appeared to sense my proximity. Up from his pale brown front end came one of his eyes. Nothing happened for a while, so I moved a little closer, and up came the other eye.

A photo of a slug usually shows her (I don't like to call an animal "it") with what looks like a V shape of two tiny horns made of the same material as the rest of her body. These come up from her front end and carry her eyes. At the tip of each horn is an eyespot. Not all slug species have these eyespots, but these slugs did, complete with lenses with which they see objects as we do but not in color. If a slug doesn't feel the need to look at something, she pulls her eyes back down into her body, where they disappear without a trace.

If eyes suggest a forehead, then two other sense organs come out of her chin—two much smaller horns that inform her of taste and scent. The sense of touch is in her entire body. She has no sense of sound but perhaps can feel vibrations. The slug I watched didn't put out his taste and scent organs, perhaps because he wasn't overly concerned. He even withdrew one of his eyes and left only the other to keep track of me.

We tend to see the slug types as disgusting. Surely they're damaging our flower beds or doing something else that's awful so we stamp on them or even put out poisons. But any life-form is fascinating if you watch it long enough, and the small, slow-moving animals are particularly favorable because you can watch any one of them indefinitely and note all kinds of behaviors. Just think what you could learn by watching slugs—not when you're high above them on a rain dampened lawn, but at eye level beside a glass aquarium planted with moss and other vegetation, complete with their favorite foods and easy access to water. I'd learn from them and find out if they'd learn from me. If a single-cell organism can learn and remember things, as has been demonstrated with paramecia, and if birds and fish can recognize individual humans (most humans can't recognize individual birds or fish), who knows what a slug can achieve?

A marvelous book, mentioned earlier in another essay, was written on this subject, *The Sound of a Wild Snail Eating* by Elisabeth Tova Bailey. It's about a snail while this essay is about a slug, but they're the same thing, really—a snail is a slug with a shell. I can't get enough of this book—it opens a world of utterly fabulous creatures that most of us know nothing about. After you read it, you will never again see a snail or a slug without marveling. You might even want one as a pet.

Worms:
Lowly and Exalted

— *Sy* —

*F*ound only in a critically endangered ecosystem known as the Palouse prairie, a storied giant was long thought to be extinct. Only a handful of sightings have been reported since the 1970s. Today there are only ten of these animals in captivity in the world. Seeing a rare species is one of the highlights of a naturalist's life—and on one day in Moscow, Idaho, I was thrilled to have the opportunity to see one.

In the second-floor laboratory at the University of Idaho's College of Agricultural and Life Sciences, PhD candidate Chris Baugher did the honors. From a plastic Tupperware container the size of a shoe box, and onto some moistened white filter paper, he dumped out several cups of black dirt. And there it was: a worm.

Not just any worm, mind you. This was a giant Palouse earthworm—portrayed in the media as a "spineless, subterranean Bigfoot," described as "Moby Worm," and considered by worm experts to be the holy grail of North American earthworms. I had read it was white, grew to grow more than a yard long, and spat saliva that smelled like lilies.

The worm before us was none of the above. It wasn't white at all—mostly reddish purple with a handsome, peach-colored forward section. It was only about eight inches long. And Baugher and soil scientist Jodi Johnson-Maynard, considered the world's top experts on the animal, admit they've never been able to detect its scent.

To my untrained eye it looked a lot like the common night crawlers they sell at the Hancock Market here in New Hampshire to bait anglers' hooks.

But night crawlers—the reddish gray species you find on sidewalks after a rain—are, like most earthworms now found in the United States, an invasive species. They arrived in ballast used to steady early ships from Europe. "Of the six thousand species of earthworms," explained Baugher, "very few are native. It may be that the giant Palouse earthworm has been here for a very long time."

Shockingly little is known about any of our native earthworms. There is only one working earthworm taxonomist in America. International earthworm experts gather at a symposium only once every four years. The giant Palouse earthworm illustrates just how mysterious are the lives of the little creatures who live under our feet—animals to whom we give little thought.

But Johnson-Maynard reminds us that earthworms have profound effects on our lives. "To many people the soil is just a black box we walk on," she says. But it's the foundation of our food chain and, she points out, importantly regulates gas exchange with the atmosphere. Soil sequesters three times as much carbon as the atmosphere, adds Baugher. And earthworms are soil's stewards.

It's difficult to learn about animals who live underground. Baugher and Johnson-Maynard have made plaster casts of their burrows. They have tried digging the worms up, but that's a good way to cut them in half—not a good thing to do to a rare species. Genetic expert Lisette Waits is working on ways to identify their worms' burrows by DNA gathered from swabbing mucus (which they secrete to speed their passage) from burrow walls.

"To cultivate the giant Palouse earthworm is a real chore," said Johnson-Maynard. Nobody is sure what type of soil it prefers, how wet to keep it, or even what it eats. Night crawlers come to the surface at night and carry leaf litter down to their burrows to feed. Maybe the giant Palouse does the same; maybe not. "We're just trying to keep them alive." (That's why the worm I saw was dumped out of its container; the researchers need to make sure their animals are still alive.)

Most of the specimens in captivity were brought in by one man, Cass Davis. He's a self-described "liberal redneck," an Earth First! environmentalist who feeds himself by hunting and fishing. "I'm quite familiar with worms," he told me. "I've put a lot of worms on hooks." He used to swallow night crawlers on a dare; that way he earned chewing tobacco as a teen. Now fifty-two, he found his first giant Palouse earthworm in 2012 in a rut on a road. It had been run over, but even in this condition, it didn't look like a night crawler. He brought it in to the university lab—and sure enough, it was the storied worm. He has a photo of it—and all the others he's found—on his cell phone. "They have beautiful lips!" he told me as he displayed the picture.

Davis is one many citizens of this corner of Idaho, including a number of farmers who have collaborated with the university scientists, who are proud to share the home of the giant Palouse earthworm. (Though some farmers—ironically the very recipients of the worms' hard work aerating the soil!—fear that if conservationists get the worm endangered status, it could restrict use of their land.) "Citizen scientists have been very important to the project," says Johnson-Maynard. Folks bring animals into the lab all the time, hoping they've found the elusive worm. One person brought them a very small snake; another brought in a leech; and once someone brought a photo of a long white thing that turned out to be the intestine of a large mammal.

Still, Baugher and Johnson-Maynard are grateful to them all. They love it that the giant Palouse gets people excited about earthworms. "It's unique to this region. It draws them in," says Johnson-Maynard. "And it really is a beautiful animal!"

Amphibians

— *Liz* —

The only amphibians we now know are the salamander and frog types (toads are frogs), and we think of them as nothing much, not realizing that tiny though they are, their brains are fully functional, with consciousness, memory, thoughts, and even emotions. In short, their brains are more or less like ours except specialized for different problems and focused in different directions. It's true that an amphibian has yet to write a book or design a nuclear reactor, but we humans have yet to successfully spend our childhoods as fish before morphing into land-based life-forms in an entirely different ecosystem with all-new problems we must then solve. That amphibians do this perfectly we see as nothing. That's just us, though, with the normal human reaction to a nonhuman life-form.

Amphibians are faced with extinction. Believe it or not, this is due in part to a certain fungus that during its infancy swims like a fish. You read that right—the spores of certain fungi swim like fish. They also swim with a goal in mind, and the goal is an amphibian to whom they can attach, just as a coral polyp attaches to a rock.

While swimming around, these fungi find their victims by sensing the proteins and sugars in their skins and swim toward them, then attach to them, grow their hyphae (their "roots," so to speak) down through the victim's skin, then suck the juices from the victim, who eventually dies of heart failure. I'd say this fungus has some powerful abilities, and those who do this are known as chytrids, which, for reasons yet undetermined (but probably through human activity), are spreading worldwide; hence all amphibians are threatened.

If amphibians go extinct, which could certainly happen, an entire taxonomic class would disappear. That's all the animals of a certain group, when normally it's just a species that goes extinct. Losing an entire class would be as bad as losing all the birds or all the mammals and worse than losing all the dinosaurs, because certain dinosaurs turned into birds and in a sense are with us now.

But despite the horror that's waiting to happen, we still don't help amphibians. We are among their most important predators, and we kill them on damp nights in the spring when they're crossing the roads. Amphibians must do this for several reasons—sometimes to find mates, sometimes to forage, and sometimes to travel to the vernal pools in the woods that form when snow melts. Because these pools are not connected to streams or ponds, fish have no way to reach them, so many amphibians lay their eggs in these pools, knowing that their eggs will not be eaten and most will survive and hatch.

Amphibians travel at night because the air has more moisture. The modern amphibians have some kind of lungs, but they also take air in through their skins, which must be damp for this to happen. Thus they must move when the air is damp and won't dry them too badly. But most are so small, relatively speaking, that no matter how quickly they move, they can't cross a road in a timely manner, which puts them at great risk. A car is to them what a bomb would be to one of us.

If not for them and their brave journey with the hardships they endured when leaving the water for the land, we would still be fish, not people with cars to squash them. Thus we owe them the courtesy of not running over them. So please, everyone who reads this, know that a careless or uncaring driver can kill as many as twenty amphibians all at once. Not even the little dinosaurs who turned into birds had to withstand that kind of predation.

Here in New Hampshire where I live, groups of volunteers go to the most dangerous roadsides—those that amphibians have been known to favor—to assist the little frogs and others who are trying to cross. The volunteers wave at the cars to ask them to slow down, and many do. Sometimes the volunteers have signs that say "Save the frogs." Sometimes the drivers wave at the volunteers to thank them. And the frogs that were hopping across the road finish their journey, hopefully to live, find a mate, and make more little frogs.

Bumble, Bee Happy, Bee Smart, Bee Safe!

— Sy —

Like many kids, I used to lie in the grass and watch bumble-bees harvesting pollen from the clover. Unlike many other bees, these big, furry bees were so gentle I could let them crawl across my palm without fear of getting stung. I loved to select one bumble to follow around the yard as she buzzed from clover to rose, rose to snapdragon, sipping nectar from her pointy black tongue and collecting pollen on her fuzzy black-and-yellow coat, later to groom it off to carry home in "pollen baskets" on her rear legs. Though our parents may have chuckled benignly at our beliefs, I'm sure lots of kids, like me, reported that our insect friends were cheerful and smart.

As it turns out, we were right.

Two separate studies published just a few weeks apart report that bumblebees show emotions, solve problems, and will teach others to how to solve problems, too.

"Even insects express anger, terror, jealousy, and love," Charles Darwin wrote in *The Expression of the Emotions in Man and*

Animals. But in the ensuing 150 years, his views on emotion fell so out of favor that few scientists even tried to look for thinking or feeling in tiny, invertebrate animals—until now.

It's difficult to study something as private as an emotion in a creature as different from us as a bumblebee. Unlike our dogs and cats, they have little reason to communicate with us. So neuroethologist Clint Perry at Queen Mary University of London came up with an ingenious experiment, published in the journal *Science.* His research team trained twenty-four bees to enter a plastic tunnel when a treat was promised. When marked with a blue card, the end of the tunnel offered tasty sugar water. A green sign meant none.

But what about an aquamarine sign? Like you or I might be, the bumbles were confused. Was it green or was it blue? They wandered around, not knowing what to do. Then the researchers gave half the bees a dose of cheer: a surprise treat of a drop of sugar water. Like people in a good mood, they then became more optimistic: They entered the ambiguous tunnel, encouraged to hope for the best. Those who had no sugar water spent just as much energy dithering around but didn't take a chance that the tunnel would be a good bet. This suggests that the metabolic effects of the sugar weren't responsible for the bees' behavior—but the boost in mood was. Who doesn't cheer up after a sweet treat?

And here's the clincher: The effect of the sugar water disappeared when the bees were given a drug that blocks the receptors for the natural brain chemical dopamine—the neurotransmitter that is associated with pleasure and motivation in humans and animals. Bees possess the same neurotransmitters as humans do. Why shouldn't they have similar emotions?

In another experiment at the same university and published in the online scientific journal *PLoS ONE*, researcher Lars Chittka watched bumblebees figure out how to use string to retrieve a snack. Artificial flowers filled with sugar water were placed under Plexiglas and tied to a string sticking out from under the plastic.

Eventually a particularly insightful bee would figure out that by yanking the string with his (or her) front legs, he could retrieve the flower and sip the liquid.

What happened when Chittka allowed unsuccessful bees to watch the insightful one was even more exciting. Most of them learned the new behavior by watching—and when these student bees were introduced to new colonies who had never seen the string-pulling technique, the learned behavior spread from bee to bee. Soon almost everyone was pulling the strings to get to the sugar water.

Sadly, just as scientists and adults are starting to appreciate the fuzzy, gentle bees we loved as children, bumblebee populations are crashing across the United States and Europe. In March 2017, the US Fish and Wildlife Service listed the rusty patched bumblebee, once common throughout the Northeast and Midwest, as federally endangered. It has lost 95 percent of its population since 1990 and now only lives in twelve states, including Maine and Massachusetts.

The rusty patched is not the only one of the 50 species of bumblebee in the United States and 250 species of bumblebee worldwide to be threatened—at least 4 others are in serious decline. (The rusty patched is just the one whose decline is best documented.) Pesticides, global climate change, loss of favorite wild food plants, and a fungal gut disease imported from Europe are all implicated.

Invertebrate ecologist Timothy Hatten, a consultant and adjunct faculty member at the University of Idaho, is among the researchers trying to help save the bumblebees, but his findings have only deepened the mystery. His team surveyed bumblebees along Canadian highways of British Columbia and the Yukon Territory and found many bees carrying the European fungus called *Nosema bombi*. Yet mysteriously, despite the infection, in these northerly locations the bees seemed to be doing fine.

Bumblebees deserve our protection. "Bumblebees are among the most important of all pollinators," says Hatten. Unlike honeybees (who are mostly Italian imports), our native bumblebees forage throughout the growing season, and their thick pile coats render them especially well adapted to cool climes like those in New England. If we lose them, we forfeit an essential link in our food chain.

But perhaps just as tragic would be the loss of a childhood icon—a friendly, fuzzy bee who enticed so many of our young selves to explore and observe the natural world.

Water Bears

— *Liz* —

*I*f one writes about black bears, as I have sometimes done, one should also write about water bears—animals that look something like black bears but are smaller. A black bear can be eight feet long and weigh five hundred pounds, but the biggest water bear is only 0.02 inch long (if that) and could almost be said to weigh nothing. No weight has been established for any water bear as far as I know. It's hard to imagine how to weigh them. You can't even see them. If your eyes are exceptionally good, you might observe one as the tiniest possible dot, but most people need a good microscope to spot them.

I've had a lifetime fascination with water bears, ever since I was in college and my dad lent me a binocular microscope so I could watch the life of a nearby swamp. One day I was watching some algae floating around when suddenly a monster appeared and charged straight at me. Shocked, I threw myself backward. My chair tipped over, and I fell on the floor. The monster was a water bear, something I'd never seen before or heard of. But when I remembered I'd observed it through a microscope, I knew it was too small to be dangerous, so my courage returned. I went back to the microscope and watched it for hours as it moved through the drop of water.

Through its transparent "skin" I could see that it was eating algae, and toward its hindquarters I saw some roundish shapes that I thought could be eggs. The possibility was good enough for me—I decided she was a girl water bear. I put her back in my jar of swamp water but was concerned about her well-being and that of her infants in the eggs (if they were eggs) perhaps soon to be born, so later I returned her to the swamp.

Later I learned she was almost certainly a female and the round shapes at her rear were almost certainly eggs, but she wasn't going to lay them. They would fall away from her with her "skin" when she shed it as a snake would do. So she wasn't as bearlike as her name suggests. But she walked with her head low and her "shoulders" high, so she resembled a bear, or so thought the scientist who named these animals.

They're also known as tardigrades, which evidently means "slow walkers." Otherwise they look vaguely like caterpillars with their segmented bodies. But instead of six legs and ten prolegs like most caterpillars, water bears have eight legs, each leg with four toes, each toe with a claw. Their little heads have tiny snouts and two eyespots. They have sensory whiskers on their bodies. They also have brains so they probably think, but what they think about is hard to imagine, as their world is so different from ours.

These are animals? Yes, and of all the animals in the world including us, they are certainly the most successful. Interestingly, most of them are female. Males exist but aren't needed for reproduction, although they do offer genetic variety now and then. Otherwise a female reproduces without a male, and her offspring are females. Water bears have been on earth for more than 500 million years (we humans have been here for 200,000 years), they now have more than a thousand surviving species (we great apes have five surviving species, four of which are in decline), and depending on what species the water bears belong to, they are found in every imaginable environment—from almost absolute zero to far above the boiling point of water, in the deepest parts of the oceans, on the highest mountains, and everywhere in between. They favor moist environments such as swamps and damp moss, but they can also live in outer space where some water bears, probably all females, were taken in 2007 as part of a biological experiment. There they were exposed to vacuum conditions (meaning no atmospheric pressure) and intense ultraviolet radiation, only to recover and reproduce successfully because vacuum conditions don't seem to bother them and they can withstand literally one thousand times more radiation than we can.

If a water bear is injured she folds herself up until she recovers, but if conditions are very bad—no moisture, maybe—she pulls her head and legs inside her body, squeezing out the water that was in there, so that she vaguely resembles a microscopic mouse dropping. In that condition she can live for many years (some say for a hundred years) while waiting for improvements. The condition she's in is called cryptobiosis, meaning "hidden life," and the form she assumes is called a tun, because whoever called her that thought she looked like a tiny wine barrel.

We think of ourselves as the ultimate species. We've spent the last eight thousand years forcing the natural world to meet our needs while water bears, in contrast, made the necessary adjustments. They hit bingo 500 million years ago, then ignored all the ice ages, droughts, and extinction periods and are here today to show what is meant by successful evolution. If we should cause a mass extinction with our wars and pollutions, a few female water bears could repopulate the planet as a dominant species and the planet would be in better hands—or if not actually hands, then at least with better toes with claws.

PART SIX

Animal Abilities

*W*hen we consider the abilities of other species, we often use them to define the animal. The first nonhuman ability that usually comes to mind is a dog's sense of smell. We all know about it, and many of us bring it up when dogs are under discussion. Or anyway I always do, mentioning a certain dog used by police for tracking suspects. Once when the search for a suspect was in progress, the dog not only directed the police down an interstate highway but knew which exit the suspect had taken—all this while riding in a police car going sixty miles an hour. Thanks to the dog, the suspect was found and arrested. Yes, we put a man on the moon, but that dog accomplished something equally surprising and without heavy machinery and a team of scientists to help.

As far as the animal is concerned, its own use of such abilities is equally important but not always on our radar. Here I remember a bomb-sniffing dog I met in an airport. He would slowly sniff a passenger until he knew no bomb was present and then would move on to examine the next person. This constant sniffing, of course, was important for air-travel safety, but the findings were of little interest to the dog himself. Or not until he came to me. Although I'd never owned a bomb, let alone transported one, he took at least three times as long to examine me as he did anyone else.

At the other end of the leash, the security guard seemed puzzled and began to appear suspicious. So I believe that the dog and I were the only ones who knew what he was doing—without changing his expression or deportment, he was learning about my dogs, but in a regular, businesslike, unrevealing manner. When he'd learned I had three dogs, a male and two females, and that none of them were present but all were in good health, instead of sitting down in front of me as he would if he'd found a bomb, he moved on to the next passenger as if he'd seen no reason not to answer his own questions as well as the airline's. He didn't lift his leg and mark my shoe to send my dogs a message—an act traditional to dogs when they've finished their olfactory investigations—but he did give my shoe a backward glance as if he'd thought about it.

Sometimes we see human abilities in other animals. If we do, we don't marvel that the animal has some of ours as well as his own; instead we judge the animal as if he were a person. An August 28, 2015, article about Koko the gorilla in *The Atlantic* said, among other things, that Koko had "the ability of a three-year-old child." The comparison of animals with human children has been applied to virtually every animal who exhibits some human ability. And it's normally a three-year-old child, rather than, say, a four-year-old child, so one must assume that those who offer this statistic are getting it from one another.

If a rabbit was judged by a weasel's abilities, would that tell us something? Could Einstein, Lincoln, or Shakespeare have found a culprit by sniffing the air as they sped down a highway? Most animals have abilities that those gentlemen seriously lacked, so if one measures the abilities of one species by comparing them with another, a dog might credit those above-mentioned gentlemen with the abilities of a very young pup.

If an animal can travel through the air at night, navigating by the stars and the sound of the ocean to find an exact spot hundreds of miles away, we tend to dismiss it as "instinct" and leave it at that. Yet it's hard to imagine the equipment—the aircraft and all it requires plus the maps or the GPS systems—that we would require if we were to do the same. What it must be like to jump up in the air and somehow find a distant spot unaided is beyond our understanding.

There was a time when we understood other species more deeply. Alas, that was thousands of years ago when we lived as hunter-gatherers. Far more work and attention were needed to obtain this understanding than most of us, consumed as we are with all our little gadgets, are willing to use today. Here I'll use information obtained from the San of the southern African savanna, who until as recently as the 1960s lived by hunting and gathering and had little contact with the "developed" world. These were the first people. They gave rise to the rest of us. And like most other vertebrates who lived in the old way, they understood the abilities of those with whom they shared the environment. Thus their name for themselves, which means "people," seems to be a species designation.

They knew what we no longer know, that the animals around them were not unlike themselves, with abilities that required serious attention. They regarded other species as those species regarded them, knowing one another through empathetic study, watching their habits and reactions and assuming (correctly) that they had plans, imaginations, memories, and emotions much as we do, realizing that we have much the same feelings and communicate with similar signals if in somewhat different forms.

As for observation, consider this example: One morning after my field was mowed, a young doe, born that spring, came out of the woods to investigate. *What happened here?* her cautious manner said. She explored as carefully as a scientist, examining the now-short grass, and perhaps from frustration because she couldn't figure out what had happened to it, she ran in a big circle and came back to try again.

As for imagination, Sy's dog Thurber barks at a bone and dances around it. He knows perfectly well it isn't animate—he's just pretending.

And as for understanding, during the time my husband was seriously ill and in bed, I'd sit beside him holding his hand. One night as his health was declining, his cat jumped up on his bed, put his paw on top of ours, and held it there, quietly joining the three of us together.

After knowing our fellow animals for 200,000 years, we humans lost touch with them. We drew the haughty conclusion that God looks like us and wants us to dominate the world. Both concepts are unlikely, but the latter is no different from the concept held by a large group of lions who once surrounded a tent harboring a few unfortunate tourists. The lions engaged in polyphonic roaring, terrifying the tourists. Many things are frightening, but nothing is more frightening than the polyphonic roaring of lions, especially if they're near, as the continuous blast of multiple deafening voices can last for twenty minutes that seem like twenty hours. Your eyes fly wide, your skin prickles, your teeth chatter, you can't hear yourself think. The lions are telling you that the place where you pitched your tent is theirs and you must go somewhere else. And you do, as fast as you can. The tourists did, so in this case, lions were dominating their part of the world; they knew very well what their noise would accomplish, and they did what they thought was right.

The journalist Roc Morin, who wrote the article about Koko, adds to this. Comparison with a three-year-old child wasn't his idea—he was just quoting—and the ending of his fascinating article says it all. "I thought of all the radio equipment and telescopes perpetually aimed at the sky," he wrote, "scanning the heavens for the faintest glimmer of intelligent life. All this, while we are still so far from truly understanding the intelligent life here at home."

—*Liz*

Abandoned Acrobats

— Liz —

One day not long ago, Sy learned that a show called Stunt Dog Experience would perform in a nearby city and immediately bought two tickets, one for her and one for me. The day of the event, the theater was jammed, but our seats were good and we watched in wonder as the dogs did things we wouldn't have imagined possible. One dog, not a big dog, jumped over a barrier that was 5 feet 8 inches tall, almost from a standing position, which for a person would be like suddenly leaping into the air to clear a barrier about 20 feet tall. Another dog stood on his hind legs to skip rope, keeping it up for what seemed like forever. Then two dogs together, both on their hind legs, skipped the same rope. Dogs ran through tunnels and zigzagged through barriers to catch a Frisbee that was already in the air, or they walked on their front legs, crawled on their stomachs frontward and backward, then jumped to their trainers' shoulders where they stood up straight on their hind legs, their front paws waving.

Many of these dogs were Australian cattle dogs or border collies, or at least had a parent who was such, but one little dog, perhaps part Chihuahua, jumped to her trainer's arms and then to his shoulder, at which point the trainer held one hand up high. The little dog jumped to the palm of his hand and stood on her front legs, six feet or more above the stage, holding her position for what seemed a long time in perfect confidence. We in the audience cheered so loudly we could have been heard for blocks.

We all know that dogs can do tricks, but *those* tricks? We'd never even imagined dogs doing anything like what those dogs were doing. But what fascinated Sy and me the most, since we're both dog lovers, were the dogs' facial expressions and body language. They loved what they were doing! A dog who was supposed to catch a Frisbee—to be thrown away from him, not toward him—stood waiting at full attention with a tense little smile, his eyes wide, his ears up, his teeth chattering with anticipation, his body poised to run, until the Frisbee flew with him right behind it. He ran faster than the Frisbee and caught it, then rushed with it to his trainer in delight.

All the dogs would wait on high alert for their turn to do a trick, smiling and quivering with excitement because they knew they could do what was expected and couldn't wait to start. After a successful trick, the dog would jump into her trainer's arms and kiss him. She would receive a tiny snack about the size of a kibble as a reward, but she hadn't done her trick so she could eat a kibble, she had done it because she could, because she loved her trainer, because she knew she was doing something very important—however extraordinary, however difficult—and found joy in her success.

Who wouldn't want a dog who might accomplish so much, who might do what seems impossible with such skill and joy and grace? Here's the answer: the millions of people who every year discard roughly 3.5 million dogs. Maybe their new apartment doesn't allow pets, or maybe a boyfriend or girlfriend doesn't like animals, or maybe they find their dog to be a nuisance, barking at noises or needing to be walked and provided now and then with food and water. The best of these millions took their dogs to a shelter—not necessarily to a no-kill shelter—and the worst of them simply turned a dog loose somewhere, maybe driving her out to the country and pushing her out of the car.

Virtually any dog can be trained to do wonderful things, wants to do wonderful things, wants to delight his or her owner, wants to do his or her best, yet abandonment had been the fate of each and every one of those marvelous dogs that Sy and I watched with such wonder. The trainers rescued many of them from pounds or shelters, some just before they were about to be euthanized. Others were simply found lost and wandering, such as one little female who was discovered as a puppy, frightened and alone on the streets of Kansas City.

These fabulous dogs had been misunderstood. I think of the little dog who, on her front feet with her hind legs in the air, stood on the upheld palm of her trainer's hand. She was probably the first dog ever to do this seemingly impossible trick, but her former owner had tossed her out, much as we might toss unwanted clothing in the Salvation Army box with no concern about who, if anyone, would use the clothing next.

Once rescued, all these dogs learned to do things no dog had ever done before. Their ability astonished us. Their delight in performing overwhelmed us. When given a chance at happiness, these dogs rose from the ashes and reached for the impossible in concert with the humans who rescued them—a kind of ressurection that most us of thought was only for humans.

What's in a Name?

— *Sy* —

O n spring days, when normal people are listening to NPR or CDs or music from their iPods, I listen to the baby monitor.

We don't have a baby; I got the device to tune in to the conversations among our chickens.

I bought it to listen for distress calls. Being a chicken in rural New Hampshire is risky business, thanks to foxes, coyotes, dogs, hawks, and a host of other predators. Our flock well knows this, and if anyone sees a predator, they call out to warn the others—and thanks to the baby monitor, their calls also summon me to the rescue.

They don't yell out "predator." Their calls are quite specific. Alarm calls announce not only the species of predator spotted but also its speed, size, and direction.

At Macquarie University in Australia, psychology professor Chris Evans and his wife, Linda, identified nearly thirty phrases the birds use to convey information to others in the flock. For instance, roosters enticing hens to food tell them when the food is especially tasty. The Evanses showed that a rooster calls at a faster rate when alerting his flock to their favorite food, corn, than when he is merely pointing out the discovery of their regular layer-mash ration.

But food and predators aren't all that chickens talk about. They may also be talking about us.

At least they are at Melissa Caughey's coop. She's the author of *A Kid's Guide to Keeping Chickens*, which won a Science Book and Film Prize from the American Association for the Advancement of Science. A science award for a book about chicken keeping? You bet. Observing her flock of ten at her chicken compound on Cape Cod, Caughey is conducting serious science—and making important new findings. She told me about her latest discovery when we met in Washington, DC, a couple of years ago, and it took my breath away.

Her flock has come up with a name for her.

"I was out there one morning, throwing scratch into the run," she told me, when she noticed her eldest, six-year-old Oyster Cracker, "was talking to me in a different voice, one I'd never heard before." Oyster Cracker wasn't just uttering the greeting "Brup? Brup?"—a chicken hello. She certainly wasn't saying, "Bwah, bwah, bwah." (That's chicken for "I'm about to lay an egg." But Oyster Cracker had long since entered what Caughey calls "henopause.") Oyster Cracker was clearly saying, with increasing emphasis and tempo until the final, higher note, like trumpet fanfare, "ba-Ba-BA-BAA!"

"It was quite regal sounding," Caughey said, "almost·like announcing the arrival of the queen!" And then Caughey noticed other hens would say it, too—only when they first caught sight of her. Hence her conclusion: "When they see me, they call my name."

This is not the first time observers have documented animals using particular sounds to refer to approaching people. Arizona professor Con Slobodchikoff has documented that prairie dogs—gregarious ground squirrels—use specific sounds to communicate to others that a human has been spotted. (They also discuss the dangers of cats, badgers, hawks, and ferrets, as well as announce the comforting presence of harmless species like cows and prong-horns, who signal safety.) Not only that, but the little mammals can communicate what color shirt the human is wearing, whether he is tall or short, and even whether the human is carrying a gun! Matching sonograms of what the prairie dogs are saying with videos of their responses to different stimuli, Slobodchikoff was even able to discover that the chatty squirrels will invent new "words" on the spot to describe objects they haven't seen before, like circles and triangles.

That's different from endowing any specific individual with an actual name—like "Sue" or, for that matter, "ba-Ba-BA-BAA!" But other animals are known to use individual names. Scottish researchers at the University of St. Andrews announced in 2000 that dolphins have names for one another, which the researchers called "signature whistles." Further studies in captivity in South Africa and in Florida, among other places, proved not only that dolphins invent names for themselves and others but that they will call out the names of loved ones when they are separated, just like you call your kids or friends when you are looking for them.

Big-brained mammals like dolphins aren't the only ones to use names. In 2008 scientist Karl Berg discovered that wild parrots do, too. Green-rumped parrotlets of Venezuela use specific peeps to identify themselves and others. They can, in effect, call out to other parrots: "Hey, Jill, it's Tom! Wanna go get some fruit?" Further-more, Berg showed how the parrotlets get their names: Like us, their parents name them.

That parrots have names in the wild shouldn't be too much of a surprise. Parrots easily learn our language, for goodness' sake; why shouldn't they have their own? Everyone knows how smart dolphins are. As for prairie dogs, at least they're mammals like us. But chickens? Too many people dismiss them as stupid.

That's a big mistake. Chickens—like most animals—are much smarter than we give them credit for. Scientific experiments show they easily recognize the faces of at least a hundred different individual chickens; they remember the past and anticipate the future; they have excellent spatial memories.

Caughey believes that my flock may have a name for me, too—and that our hens probably have names for each other as well as for their humans. We wonder what other species do this: elephants, wolves, crows? What about fish? In coming years, we predict, we're likely to hear similar discoveries about these animals.

"Animals will share their wisdom," Caughey promised. "But you have to listen."

As for me, I'm staying tuned to the baby monitor. I'll let you know what I find out.

In the Snow

— *Liz* —

The amount of snow we have at times is hard on many animals, especially those who can't store food or hibernate and those with short legs who must plow through the snow, forcing and stumbling as we do if we don't use snowshoes. Such winters are hard for bobcats, coyotes, foxes, fishers, and the like, although lynxes with their wide feet do better.

Deer suffer despite their long legs and their ability to bound through snowdrifts. This takes much energy and spends many calories, of which deer in the winter find few. Their digestive systems go into winter mode, which helps them but won't always sustain them. Some are overcome by weakness and they starve; when the snow melts we find their carcasses, usually a few bones in a circle of fur, the deer's winter coat that a scavenging predator has torn away. Mother Nature isn't always kind, but she's always effective. If not for carcasses, many of the carnivores would starve.

As I've mentioned earlier, I feed about twelve deer who live in the woods near my house. People are not supposed to feed deer. Even so, it's my land and I'll put corn on it three times a day if I want to—the benefits and problems and my reasons for doing so are described in an earlier essay. By April the deer are still doing well, and the fawns they bear later are, too, and they all come the next winter for more.

Who else does well in a bad winter? Bears do if they've eaten enough to tide them through their hibernation. Overwintering birds do if they live near someone with a feeder. Squirrels and chipmunks do if no other animal steals the acorns they have stored for that purpose. Beavers do because they've brought branches to the ponds where they've built their houses and stuck the branches into the mud on the bottom of the ponds. In winter the beavers swim under the ice to bring branches back to their houses. Mice often live with them in their houses, as these houses are built near the shores of the ponds where mice can access them, just as they access our houses. I'm not sure what the mice eat. Perhaps they, too, have stored food, or perhaps the beavers drop scraps while they're chewing.

But even without the use of someone else's shelter, mice and voles do very well in winter. Not only do they cache enough food for the winter in different places but they make tunnels under the snow so they can reach their caches. And in these tunnels, predators can't find them.

What about pets? My Chihuahua, mentioned in earlier essays, was a rescue dog, born in a southern city and raised in an apartment. He hated the cold, rain, and snow, and he stayed in the house using puppy pads for bowel and bladder relief and sleeping right next to me in bed under the covers, with never a thought about the outdoors. But then a big dog came to visit us. She's a young Labradoodle, she loves the snow, and she was with us when I wrote this because her owners were away. She had impressed the Chihuahua, who wanted to go wherever she went, so he would follow her outdoors. In February, after a storm called the "storm of the century," snow had piled up to our roof, so the Labradoodle would run up the pile and stand on the roof. She very much liked this because she could see for miles. The tiny Chihuahua didn't quite dare follow, but he was thrilled by the sight of her up so high. When she'd come down from the roof, they'd visit the deer-feeding station, sniffing every footprint that every deer had left in the snow. This thrilled them both.

One day a deer was there, but on the far side of a snowbank. The snowbank was so high she didn't see the dogs coming, but when they appeared, she bolted and they chased her. The little dog's legs were so short that normally he'd sink to his belly with each step he tried to take, but the snow had a crust that supported him. I called the dogs back and the big dog came, but the tiny dog kept going. He normally does what I ask, but with all the snow and an enormous creature trying to escape from him, he felt wild. To chase a deer is a serious misdemeanor known as "running deer." If a dog is caught at it, he can be destroyed, so I shouted. I was also afraid that the deer would turn on him and do him harm or even kill him, but on they ran until the deer was in the woods. Feeling he had done his job, he calmly trotted back. Who would dream that a tiny dog from an apartment in a southern city could be stimulated by heavy snow and another dog, however big, and wind up running deer?

Music for Animals

— *Sy* —

*O*ften as my border collie and I drive to our favorite hiking areas and to playdates, we garner stares from pedestrians. Even when my car windows are tightly closed, it's pretty obvious to whomever we pass that Thurber and I are howling together at the top of our lungs.

Appropriately, Thurber seems particularly inspired by the soaring notes of the song "Say Something" by A Great Big World. I used to sing along; but when Thurber joined in, I switched to howling. I love singing and particularly enjoy doing it with my dog. But recently I worried: Is he howling because he likes it, or because he can't stand my soprano?

I consulted an expert on animals and music, University of Wisconsin–Madison Professor Emeritus Charles Snowdon, to find out. Do animals enjoy music, like we do? If so, what kind do they like?

Psychologist Snowdon first became interested in animals and music in 2008 when University of Maryland composer and cellist David Teie contacted him with an intriguing question: How does music affect us emotionally?

Teie realized he couldn't answer his question using human subjects because all humans have already heard music, and their emotional response to any given piece could be associated with millions of other learned factors. But Snowdon had a captive colony of little monkeys called cotton-top tamarins, whose elaborate communication system he studied. None of the monkeys had ever heard music. Maybe they could be good test subjects.

Snowdon was intrigued. Previous studies of animals and music had been inconclusive. Many people swear their animals enjoy the same music they do. Farmers keep barn radios tuned to classical music to calm pigs and cows. Some owners claim their dogs like heavy metal music; others say their dogs love classical.

A 2013 Japanese study showed that goldfish could hear the difference between Bach's Toccata and Fugue in D Minor and Stravinsky's *Rite of Spring*. The fish quickly learned to bite on a red bead when they heard one composition but not when they heard the other. But did they like the music? And if so, how would we know?

Sensibly, Snowdon points out that animals, like people, approach what they like and ignore or retreat from what they don't. But at home it's hard to tell for sure if your dog likes heavy metal music. If he seems happy when you come home from work and put on your favorite CD, your dog might just be pleased to see you—or relieved that he can finally go outside to pee.

Teie and Snowdon took a novel approach. The first step, explains Snowdon, was "to evaluate music in the context of the animal's sensory system." Humans like music that uses human tones, tempi, and vocal ranges; we don't even recognize music outside these ranges as music. (Whale researchers Katy and Roger Payne had to significantly speed up the songs of humpback whales to even figure out that these long, complex compositions were songs!)

Much of our music is too high or too low, too slow or too fast, for other species to enjoy. And many animals hear different ranges than we do. When Japanese researchers played Mozart to rats, the rodents ignored frequencies below 4,000 hertz. (And we, on the other hand, can't hear their laughter—it's above our threshold of hearing.)

Based on recordings Snowdon made of tamarins in the lab, Teie composed some music specifically for these monkeys. Typically, tamarin voices are high-pitched and fast-paced. But importantly, notes Snowdon, "David and I explicitly did not replicate the monkey sounds in the music." Rather, they used the acoustic features he thought would lead to calming and arousal in general, and translated them into the frequency range and tempos that tamarins themselves use.

The tamarins responded powerfully. Tunes with soothing harmonic structure and legato notes produced calm. Those with short, sharp, dissonant notes created agitation. Teie and Snowdon published their results in 2009 in the British journal *Biology Letters*.

Lately, the duo has been working on music for cats. Cats vocalize about one octave higher than people. (Dogs' voices, like dogs' bodies, are far more variable than cats'; that's one reason they chose cats, not dogs, for their latest experiment.) Teie based one song on the tempo of a cat's purrs; the other featured a beat that mimicked the fast rhythm of a kitten suckling from its mother.

It might not sound that great to us (personally, I rather like it), but the forty-seven cats tested vastly preferred it to human classical music. When cat music played, the study subjects purred, walked toward the speaker, and some even rubbed against it. The classical music was ignored.

And what about my Thurber's howling? Snowdon told me he doesn't know for sure. But he notes that Wisconsin-based animal behaviorist Patricia McConnell has found that people of different languages as well as dogs of many breeds respond similarly to similar vocal tones: Short staccato sounds stop an action; descending slides are soothing. Wolves—dogs' ancestors—howl (as his monkeys call) to proclaim and sustain group solidarity. It may well be that Thurber and I are both having a blast, singing together the praises of our pack.

Animals Who Imbibe

— Sy —

"What's the cheapest beer you sell?" my husband asked with an air of urgency at the state liquor store.

We weren't college students planning a keg party. It wasn't New Year's Eve. The beer wasn't even for human consumption. The beer in question was for our pig.

We discovered Christopher Hogwood's fondness for beer his first summer as a piglet. My husband had been enjoying a cold one and offered Chris a taste. It turned out Christopher loved the stuff. Soon, whenever the pig saw anyone holding a bottle, he'd chase them until they surrendered and let him suck it dry. (This could prove quite intimidating once Hogwood topped 750 pounds.)

Keeping our pig in his favorite beverage, my husband became a familiar figure to the beer guys at the store. Each time he'd come in, they'd ask how much Chris weighed now, and with great interest tried to figure out how much he could drink before he became disorderly.

Even if our beer budget could have answered the question, we did not want to see a 750-pound pig with sharp tusks wandering around the yard tipsy. Christopher never got drunk. But plenty of other animals do.

Harris Center for Conservation Education Science Director Brett Amy Thelen told me about the black bear who ransacked campers' coolers and, with teeth and claws, punctured and drank thirty-six cans of Rainier beer before passing out in a tree at the Baker Lake Resort in Washington State. (He drank one can of Busch but then went back to his favorite brand.) His preference was so strong that when the fish and wildlife department had to relocate him, they baited the trap with the usual doughnuts and honey but also added—you guessed it—Rainier beer. And they caught him.

When I spoke with her, Thelen had been researching intoxication in the animal kingdom in preparation for a talk as part of the center's popular series, Nature on Tap, at our local inn. ("What better topic for a happy hour talk?" she reasoned.) Some of the stories she's heard are apocryphal: Elephants don't get drunk on rotting fruit lying on the ground. (That one was debunked by a University of Bristol study: Elephants eat the fruit off the tree before it has time to ferment.) But birds can and do get drunk from eating fermenting fruit, she told me. The problem was so severe among Bohemian waxwings feasting on rotting mountain ash berries one year that in the capital of Canada's Yukon Territory, the local environmental agency had to set up a drunk tank (actually a plastic hamster cage) where people could bring birds who needed to safely sleep it off.

Why would vulnerable birds purposely impair their nervous systems in this way? They aren't getting drunk on purpose. They're just trying to eat a lot of fruit to prepare for winter, and that winter there was a glut of berries. The birds were essentially poisoned by what looked like good, healthy food.

But interestingly, other animals seem to seek out intoxicating substances for the same sometimes complex, sometimes stupid, and occasionally edifying reasons we do. Thelen and I traded more stories. I had a few of my own.

Ever hear of courage in a bottle? Apparently the color-ful baboons known as mandrills know just where to find it. In *Animals and Psychedelics*, researcher Giorgio Samorini reports that in Gabon, a male mandrill will often prepare for combat with another male by eating the root of the iboga plant. He waits two hours for the drug to take effect before attacking his rival. "The fact that the mandrill waits like this," the author writes, demonstrates "a high level of awareness of what he is doing."

Feeling rejected by the opposite sex? Certain fruit flies respond by drowning their sorrows in booze. University of California neuroscientists allowed some male flies to mate with females, but placed others with females who had already mated so would reject new suitors. Those who were sexually rejected were four times more likely to choose food vials containing alcohol than flies who had had scored with the ladies.

And then some animals may simply be experimenting with intoxication to experience an alternate reality. A BBC team filmed groups of young bottlenose dolphins playing with a particular species of poisonous puffer fish. Each dolphin would gently squeeze the fish in his or her mouth, causing the fish to release a hallucinogenic toxin. Then the dolphin would pass the fish to a neighbor, almost like a person passing a joint. Afterward the dolphins would stare at the water's surface as if mesmerized by their reflections.

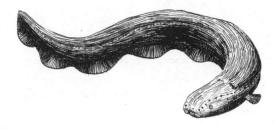

Do Animals Dream?

— Sy —

The electric eel exhibit at the New England Aquarium has a feature that makes it a favorite. Whenever the eel is hunting or stunning prey, the charge powers a voltmeter above his tank. It lights up when the eel is using his electricity, and allows you to see the invisible—like magic.

One day I saw another magical thing happen in the tank. Thanks to the voltmeter, I was able to watch the eel dream.

It happened when I was standing in front of the exhibit with Scott Dowd, the lead aquarist for the freshwater gallery, watching the eel resting motionless at the bottom of the tank.

"I think he's asleep," I said to my companion.

"Yes, that eel is catching some serious z's," he agreed.

Being hard-core fish enthusiasts, we continued to watch transfixed while the electric eel slept. And that's when it happened: A big flash shot across the voltmeter display—and another and another.

Electric eels hunt while swimming forward, wagging their heads to and fro, sending out electric signals that bounce back to them, sort of like a dolphin's echolocation. But he was still motion-less. So what was the flash for?

"I thought the eel was asleep!" I said to Dowd.

"He *is* asleep," he replied.

We realized at once what we were almost surely witnessing. The electric eel was dreaming.

"It would appear that not only do men dream," Aristotle wrote in *History of Animals*, "but horses also, and dogs, and oxen; aye, and sheep and goats. . . ." It was obvious: Like most of us, Aristotle had watched sleeping dogs twitch their ears, paddle their paws, and bark in their sleep. Surely other animals dreamed as well.

But since Aristotle's day, more "modern" thinkers denied that animals could dream. Complex and mysterious, dreams were considered the exclusive province of so-called higher minds. As brain research advanced, however, researchers were forced to concede that Aristotle was right. Animals do dream. And now we are even able to glimpse what they dream about.

Since the 1960s scientists have understood that our dreams happen during the rapid eye movement (REM) phase of the sleep cycle. During this time our muscles are normally paralyzed by the pons of the brain stem, so that we don't act out our dreams. In 1965 researchers removed the pons from the brain stems of cats. They discovered the cats would get up and walk around, move the head as if to follow prey, and pounce as if on invisible mice—all while asleep.

By 2007 we would get an even more vivid picture of animals' dreams. Massachusetts Institute of Technology scientists Matthew Wilson and graduate student Kenway Louie recorded the activity of rats' brains while the animals were running a maze. Neurons fire in distinct patterns while a rat in a maze performs particular tasks. The researchers repeatedly saw the exact same patterns reproduced while the rats slept—and they saw this so clearly they could tell what point in the maze the rat was dreaming about and whether an individual rat was running or walking in his dreams.

The rats' dreams arose from the hippocampus, the same area in the brain that seems to drive humans' dreams. It's an area known to record and store memories, and that supports the notion that one important function of dreams is to help us remember what we have learned.

Of course, it's important to a lab rat to remember the right way to run a maze. So if rats dream of running mazes, what do birds dream about? Singing.

University of Chicago professor Daniel Margoliash conducted experiments on zebra finches. Like most birds, zebra finches aren't born knowing their songs; they learn them, and young birds spend much of their days learning and rehearsing the song of their species. While awake, neurons in the forebrain known as the robustus archistrialis fire when the bird sings particular notes. The researcher was able to determine the individual notes based on the firing pattern of the neurons. While the birds were asleep, their neurons fired in the same order—as if they were singing in their dreams.

Much less work has been done on fish than on mammals and birds. No one has found REM sleep in fish—yet. But that does not mean they don't dream. Interestingly, no one has discovered REM sleep in whales, either. But whales almost surely dream. They are long-lived, social animals with very big brains much like our own, and for whom long-term memory consolidation is crucial. And if you were looking for rapid eye movement in sleeping owls, you'd never see it—because owls' eyes are fixed in their sockets. That's why they need to turn their heads around, *Exorcist*-style. Yet owls' brain waves show they dream, too.

Fish do sleep, however—that much is well known. It's been carefully documented that if zebra fish are deprived of sleep (because pesky researchers keep waking them up), they have trouble swimming the next day—just as a person would have trouble concentrating after a dreamless night.

What might an electric eel dream about? The voltmeter at the New England Aquarium showed us the answer: hunting and shocking prey.

Window to the Wild

— *Sy* —

We couldn't figure it out. Many nights a week—sometimes at midnight, sometimes 2:00 or 4:00 a.m.—our border collie, Sally, who slept all night in our bed with us, suddenly erupted into explosive barking. Since she slept with her head on our pillow, her voice was directly in our ears. Naturally we longed to discern the cause of her excitement. But her voice was so loud we couldn't hear anything else.

One cold night I got up and went outside to look around. There was a fox standing in the street, looking up at our bedroom window and screaming.

Now I understood. This was a dispute about the Smell Lane.

The Smell Lane is what I call the route of Sally's and my morning walk together. It runs through a neighbor's backyard, crossing a footbridge over Moose Brook, though a woods of ferns and hemlocks before going over a second footbridge and circling back toward home. Lots of animals use this route, and I always look for signs: the paw prints and tail drags of mice, the scales of pinecones left by red squirrels, the half-moons of deer hooves, the dino-prints and black-and-white scats of turkeys (you can even tell the sex of the individual who left it—the males' are J-shaped).

But it's Sally who knows them all. A dog's sense of smell has been estimated to be 10,000 to 200,000 times as acute as our own. Experts have likened dogs' scenting superpowers to being able to taste a half teaspoon of sugar in an Olympic swimming pool of water (per Barnard College dog researcher Alexandra Horowitz), or being able to see an object three thousand miles away (per James Walker, formerly of Sensory Research Institute). Sally stops many dozens of times on our half-hour walk to carefully smell who was there. How I wish she could tell me what she knows!

All dogs are interested in smelling the world, and some—coonhounds, search-and-rescue dogs—are professionals. But I had never met an amateur who was as interested as Sally, especially a female. It may have had to do with her past. Before she came to us, we were told, Sally came from a "bad neighborhood." Did this mean she came from a town known for crime or drug use? Low test scores in the schools? No, we were told: The neighborhood was bad because it was full of coyotes. We learned Sally had often run away from her previous owners. But she was lucky to return, because there were so many coyotes where she lived that not only cats but also dogs were known to have fallen prey to these wild canids—actually crosses between western coyotes and northern wolves, combining the adaptability and smarts of the coyote with the strength and group culture of the wolf.

For Sally, knowing all the individuals of different species in her old neighborhood may have been a matter of life and death. Hence, perhaps, her unusual preoccupation with the comings and goings along the Smell Lane. The fox screaming beneath our window was surely somebody she knew. And who knew her. Though they had never met, they had been carrying on a meaningful correspondence.

Foxes, who like wolves and dogs are canids, hold territories (one hundred acres or more); in summer they may be protecting their dens, and in winter, when their diet changes from mainly berries and insects to birds and small mammals, they are probably posting *keep out* on their hunting grounds. They mark the boundaries of their territory by spraying urine on prominent rocks and trees and, less frequently, leaving piles of their small-dog-like feces. The scent of fox urine is so strong even I could detect the sweet, skunky scent along our walk. And I had noticed that Sally made a point of marking over these *keep out* signs. (She marks more than any female of any species I have ever known.)

So this, I suspect, is what brought the fox to scream at our dog from under our bedroom window those nights: Sally had been wrecking the fox's *keep out* signs, and the fox was having none of it.

Our domestic animals can straddle two worlds: those of their human families, and those of their animal neighbors. Sometimes they can tell us about the larger, wilder world outside our walls and windows. Once, avian intelligence researcher Irene Pepperberg, now of Harvard University, took her famous talking African grey parrot, Alex, home with her from the lab. Alex had been taught to speak English meaningfully and knew hundreds of words. From his trainer's picture window, Alex looked out and saw, for the first time in his life, an owl. He began to scream, "Want to go back! Want to go back!" Alex knew, from instincts stretching back millions of years, to the ancestors of parrots and the ancestors of owls, that owls are dangerous predators. He was telling his trainer about his ancient, instinctual knowledge in a human language he learned in a twenty-first-century university laboratory.

And this is one of the many miracles of living with even the most common of pets. Once in a while they can give us a glimpse of the unseen lives of our unknown neighbors—wild animals who are near at hand but, at least by humans, are little understood.

As for our Sally, foxes no longer wake her in the night; she's gone deaf with old age, not uncommon with border collies. But she doesn't miss much. Every morning she and I still walk the Smell Lane, where, with her exquisite nose, she reads the morning mail.

Different Information

— *Liz* —

*I*t's difficult to summarize how any individual animal perceives the world, but it's extra difficult to do this for dogs in general. No other species, not even our own, can approach the dog in physical variety—a big Great Dane can be over three feet tall at the shoulder and weigh about two hundred pounds, yet it belongs to the same species as the Chihuahua, six inches at the shoulder and weighing maybe nine pounds.

So to start with vision, your height makes a difference. Just as you can see farther when up on a ladder than when standing on the ground, so can a Great Dane see farther than a Chihuahua, a fact I notice every day when outdoors with my Chihuahua, whose visual world doesn't go beyond the lawn, or even much beyond his little nostrils when we're in long grass. If you lie on your side with your head on the ground, you'll get an idea of his range. He therefore isn't informed by vision as much as are those who are taller.

How about hearing? Here he does better. When he hears a meaningful sound, he turns to face it just as we would, as if to see what made it, and he's sensitive to sound even when he's sleeping.

At night he sleeps in my bed under the covers, and if he hears a suspicious noise he leaps up barking.

Of course, dogs are best informed by scent, and as we all know their sense of smell is vastly better than ours. What's interesting about this little dog, therefore, is his realism about scent information. When he's in my office and hears something, he jumps up to look out the window, sniffing as he does. Thus he opens his eyes, his ears, and his nose to any evidence he can collect about what he's just heard.

This changes when we go outside. There the world of odors is readily available, and he runs here and there with his nose to the ground, or, because he grew up in a city and is a city dog at heart, he also goes to my car and sniffs the tires.

It makes sense if you think about it. A sight or sound can vanish in an instant, but a scent will cling. A sight or sound may be experienced quickly and tends to transmit just a tiny, very specific amount of information—often more than enough to tell you what you need to know. But it doesn't compare to a scent, as a scent remains available for a long time, and just one sniff can load you with all kinds of information.

Let's say you look out the window and think you glimpse a mountain lion in your field. You gasp and run out the door and around the house to see it better. The dog runs with you. But by then nothing is there. You, the human, see and hear nothing, the ground is dry and hard, and if in fact there was a mountain lion, it left no tracks.

For all you know, the whole thing was an illusion. But while you stand there, looking around and wondering, the dog learns that a very large cat has just passed by, a female about four years old with kittens for whom she was lactating; that she had eaten recently, not fresh meat but carrion from a carcass, probably a small animal; and that she moved her bowels not long before you saw her, so there's a scat in the woods, maybe five hundred feet from where you're standing.

You are (in this case) a mature human being with a college degree, and your dog is just a little nobody, utterly dependent on your care. But in the end, who knows more about your surroundings? You or him?

My favorite story about the power of scent involves our marvelous veterinarian, Chuck DeVinne. He's also my neighbor, and once when he was away on a trip, his dog went missing. Now and then his dogs escape, and usually they come by my house, but since I hadn't seen them I was afraid that Chuck's dog was lost. The entire neighborhood went on high alert, but our vision and hearing gave us nothing.

But Chuck knew what to do. As soon as he got home he collected the clothes from his trip that were to be laundered and bundled them into his car. He ran the motor with the heat on until the car was hot and the contents were warmed, then he opened the windows and drove slowly through the area where the dog was last seen. His scent poured from the car for the wind to carry. It clung to the bushes along the road that led to his house, and his dog found her way home quickly.

Does Anybody Know
What Time It Is?

— *Sy* —

*A*s I write this, I am preparing to leave on a two-week book tour—sadly, without our elderly border collie, Sally. She loves our shared husband, and he loves her, but as a formerly neglected rescue who now rules the roost, she prefers that both of us stay home to staff her empire. Will the two weeks we're apart feel like an eternity to her?

Some researchers say that time is an abstract notion only humans can comprehend. William Roberts, an animal cognition researcher, says animals are "stuck in time," living, as the Buddhists exhort us all to do, only in the moment.

Yet it's undeniable that animals possess their own versions of internal clocks and calendars. Some of these are obvious. Sally, like most dogs, knows when it's dinnertime and barks at 5:01 p.m. if we haven't filled her bowl. One could argue that her emptying stomach is her hourglass.

But many dog owners who work outside the home report that their dogs not only know the dinner hour but anticipate the time the owners are coming home from work each day. Some cats, too, wait by a window or door at the appointed hour. Rupert Sheldrake, the controversial English biologist, has known parrots who announce verbally, in English, when their owner is about to return. He's convinced they do it telepathically, and that it has nothing to do with timekeeping.

Another possibility, in dogs at least, is just as intriguing as telepathy: that they smell time. Dogs' sense of smell is legendary. With up to 300 million olfactory receptors in the nose (humans have a paltry 6 million), dogs can use their super-sense of scent to find bedbugs, identify diseased beehives, and locate corpses underwater.

A few years back, the BBC aired a show on which animal behaviorists investigated this question with a hound named Jazz. One of his owners, Christine, always came home at 4:00 and fed and walked Jazz. She had noticed that each day at 4:40, Jazz leapt onto the couch and stood at attention, twenty minutes before husband John came home, as if he were waiting for him. And indeed the BBC crew filmed this happening over and over for four days of the workweek. Was it possible that as John's scent decayed with time, scent molecules—not stomach or bladder contents— were the sand in Jazz's hourglass?

On Friday researchers tested the theory. That day, after work, Christine swung by John's football club and picked up some of his T-shirts, redolent with his sweat. She wafted these around the living room, spreading fresh scent—and turning back Jazz's olfactory clock. That afternoon, at 4:40, Jazz stayed dozing on the dog bed and didn't get up till he heard John open the door. Jazz rushed to greet him as if his arrival were a happy surprise.

Our pets, of course, are not the only animals who track time. Creatures as small as bees do so, as Austrian ethologist Karl von Frisch (who discovered that honeybees have color vision and communicate the location of nectar via dance) observed as early as 1953. One of his students, Ingeborg Beling, trained bees to visit a site at a given time, and found their internal timekeepers to be accurate to within fifteen minutes—even when shielded from environmental cues, in experiments with constant temperature, light, and humidity. University of London researchers found that despite constant sunlight in the Arctic summer, bumblebees there stick to a daily schedule, returning to their underground nests well before midnight.

For a food reward, a rat can be trained to press one lever every two seconds, and a different lever every eight seconds. Squirrels understand that with time, some foods spoil; that's why they dig up perishable food items first. Crows, rats, orangutans, and pygmy chimps have been shown to differentiate between now and later. The classic experiment goes like this: The animal is shown two jars, each with a treat, but one of them—say, a frozen cube of juice—disappears within a short period of time. After five minutes, the animals are allowed to choose one jar to open; they get a second chance in an hour. After just a few trials, the animals all chose the vanishing treat first, realizing they could save the other for later.

Without clocks or numbers, how do animals "count" time? Time was not the invention of humans. Time is nothing but regular cycles of motion: the tick-tock of a clock or metronome; the earth's spin; the vibrations of the cesium-133 atom (from which we get the atomic second). Time can also be measured by the decay of a scent, by daily fluctuations of the strength of the earth's magnetic field (which may be how bees do it), or by some other method we have yet to discover.

But what does time feel like to another species? Researchers can test how different species perceive, for instance, what is called critical flicker frequency—the point at which an intermittent light seems steady because we cannot process the flicker (similar to the way we don't see the image on our TVs as a bunch of pixels but as a whole image). Trinity College researchers trained animals of different species to behave one way when they see a flickering light and another way when they see a steady shine. They found that small-bodied animals with fast metabolic rates perceive more information in a unit of time. A fly, for instance, can see 250 flashes per second. Larger animals tend to experience action more slowly. A leatherback sea turtle can only see 15 flashes per second.

I pray that for Sally, time flies while we're apart. But before I leave, I'm sealing my dirty gym clothes in plastic bags. My husband can open them for her as needed.

Octopus:
The MENSA Mollusk

— Sy —

Everyone wanted to pet Octavia.

And no wonder. She was beautiful, graceful, and affectionate. The fact that she was boneless, slimy, and living in painfully cold, 47°F water deterred none of us.

What thrilled us—me, New England Aquarium volunteer Wilson Menashi, and four visitors from the environmental radio show *Living on Earth*—was the surprising fact that Octavia, who clearly wanted to be petted, was a giant Pacific octopus.

When her keeper, Bill Murphy, opened the top of her exhibit, Octavia had recognized Wilson and me immediately; we'd been working with her for several weeks. Turning red with excitement, she flowed over toward us from the far side of her tank. When we put our hands in the water, her arms rose to meet ours, embracing us with dozens of her strong, sensitive, white suckers. Occasionally Wilson handed her a fish from the plastic bucket perched on the edge of her tank.

Soon the *Living on Earth* crew joined in. People were tentative at first. In movies and stories, octopuses are portrayed as monsters, and the giant Pacific is the largest and strongest of them all. A single sucker on a large male can lift thirty pounds, and the animal has 1,600 of them. Octavia's were strong enough to leave hickeys on our arms. But she was so curious and friendly that no one could resist the chance to touch her skin, which was soft as custard. We stroked her much as we would a dog, enchanted with the spectacle of her color-changing skin, the sensation of her suckers, the acrobatics of her many arms.

Then, as Menashi reached for another capelin to feed her, we realized the bucket of fish was gone.

While no fewer than six people were watching, and three of us had our arms in her tank, Octavia had stolen the bucket right out from under us.

"Octopuses are phenomenally smart," Menashi says. And he should know: He has worked with them for twenty years and is expert in keeping these intelligent invertebrates occupied. Otherwise, they become bored. Aquariums design elaborate escape-proof lids for their octopus tanks, and still they are often thwarted. Octopuses not infrequently slip out of their exhibits and turn up in other tanks to eat the inhabitants. Many aquariums give their octopuses Legos to dismantle, jars with lids to unscrew, and Mr. Potato Heads to play with. Menashi, a retired inventor, designed a series of nesting Plexiglas cubes, each with a different lock, which Boston's octopuses quickly learned to open to get at a tasty crab inside. And just recently, aquarists at Kelly Tarlton's Sea Life Aquarium in New Zealand teamed up with Sony engineers to teach a female octopus named Rambo to press the red shutter button on a waterproofed camera to take photos of visitors, which the aquarium sells for $2 each to benefit its conservation programs. Though there's no evidence that Rambo realizes the end product of her photography, she learned to work the gadget in just three attempts.

Intelligence so like our own may seem surprising in a creature so unlike us. "Short of Martians showing up and offering themselves up to science," says neuroscientist Cliff Ragsdale of the University of Chicago, octopuses and their kin "are the only example outside of vertebrates of how to build a complex, clever brain."

The octopus brain looks very different from a human's. Our brain sits like a nut in the shell of our skull. Octopuses lack bones of any kind, and their brains wrap around the throat. Our brain is organized into four lobes. Theirs has fifty to seventy-five lobes, depending on how you count them. Most of our nerve cells are in our brain. Three-fifths of an octopus's nerve cells are in the arms.

The wonder is that octopuses and humans may think, in many ways, alike. We both enjoy learning new things, solving puzzles, meeting new friends. And possibly, we both enjoy a good joke: When Octavia stole the bucket, she didn't eat any of the fish in it. When we finally realized she had taken it, we saw she had wrapped it in the webbing between her arms, as if she was purposely hiding it from us. As long ago as the turn of the third century, Roman naturalist Claudius Aelianus wrote of the octopus that "mischief and craft are plainly seen to be the characteristics of this creature." Perhaps Octavia especially enjoyed her caper for having outwitted us humans.

"So if an octopus is this smart," one of our guests asked her keeper, "what other animals are out there that could be this smart—that we don't think of as being sentient and having personality and memories and all those things?"

An excellent question indeed.

Acknowledgments

————•————

*T*his book was born thanks to friendships. Coauthors Liz and Sy have been best friends for thirty years. The editor of this book, Joni Praded, has been Sy's friend almost as long, and was responsible for editing four of her books and one of Liz's. One essay in this book tells the remarkable experiences of another dear friend, Katy Payne, and is included thanks to her generosity. The writer of our book's preface, Vicki Croke, is also a treasured friend, and in fact was the person who suggested we write most of these stories for the *Boston Globe*, in the form of a regular column, in the first place.

For editing those columns, and for sharing them with tens of thousands of readers, we're grateful to our *Boston Globe* editors, Hayley Kaufman, Amanda Katz, and Linda Matchan. We're thankful to our literary agent, Sarah Jane Freyman, for her encouragement and support. In addition, we are grateful for the help and support of the following people and one organization: Howard Mansfield, Steve Thomas, Richard and Runi Estes, Anna Estes, Charles DeVinne, Joel Glick, Gary Galbreath, Selinda Chiquoine, Jody Simpson, Judy and Robert Oksner, Saibhung Singh and Saibhung Kaur Khalsa, Nampal Khalsa, Stephanie Thomas, Robert Kafka, and the New England Aquarium.

From three lions and a shark to a bee and a water bear, we thank the animals who appear in these essays. And for the deepest of friendships, we thank the animals who shared our lives during the years that we gathered the material and who appear in some of the essays. We owe special gratitude to Tess, Mika, Octavia, and

Thurber, to Christopher Hogwood and Sally, to Pearl, Sundog, Georgia and Sheilah, to Kafka and Čapek, Claude and Lilac, and also Coaly and Shy.

Illustration Credits

Unless noted below, all images are in the public domain.

page xi   William & Robert Chambers, *Chambers's Encyclo-paedia—A Dictionary of Universal Knowledge for the People* (Philadelphia: J. B. Lippincott & Co., 1881)

page 12   iStock.com/Andrii-Oliinyk

page 18   Worthington Hooker, *Natural History for the Use of Schools and Families* (New York: Harper & Brothers, 1882) 320

page 50   Louis Figuier, *Reptiles and Birds: A Popular Account of the Various Orders: With a Description of the Habits and Economy of the Most Interesting* (London: Cassell, Petter and Galpin, 1869): 467

page 110   Angela Boyle, angelaboyle.flyingdodostudio.com

page 121   Charles L. Stuart, *Everybody's Cyclopedia* (New York: Syndicate Publishing Company, 1912)

page 124   William Dwight Whitney, *The Century Dictionary: An Encyclopedic Lexicon of the English Language* (New York: The Century Co., 1911)

page 130   Angela Boyle, angelaboyle.flyingdodostudio.com

page 139   Buel P. Colton, *Zoology: Descriptive and Practical* (Boston: D.C. Heath & Co., 1903): 132

page 149   Charles L. Stuart, *Everybody's Cyclopedia* (New York: Syndicate Publishing Company, 1912)

page 153   Angela Boyle, angelaboyle.flyingdodostudio.com

page 168   William Dwight Whitney, *The Century Dictionary: An Encyclopedic Lexicon of the English Language* (New York: The Century Co., 1911)

page 174   S. G. Goodrich, *The Animal Kingdom Illustrated, Arranged after Its Organization: Forming a Natural History of Animals, and an Introduction to Comparative Anatomy* (New York: A. J. Johnson & Co., 1885): 154

page 177   Angela Boyle, angelaboyle.flyingdodostudio.com

page 183   Charles L. Stuart, *Everybody's Cyclopedia* (New York: Syndicate Publishing Company, 1912)

page 186   S. G. Goodrich, *The Animal Kingdom Illustrated, Arranged after Its Organization: Forming a Natural History of Animals, and an Introduction to Comparative Anatomy* (New York: A. J. Johnson & Co., 1885)

Index

Index

Index

Index

About the Authors

Saibhung Singh Khalsa

"Part Indiana Jones, part Emily Dickinson," as the *Boston Globe* describes her, Sy Montgomery is an author, naturalist, documentary scriptwriter, and radio commentator who has traveled to some of the world's most remote wildernesses for her work. She is the author of numerous award-winning books, including her memoir, *The Good Good Pig*, an international best seller; *The Soul of an Octopus*, both a best seller and a 2016 National Book Award finalist; and numerous award-winning books for children and young adults. She lives in Hancock, New Hampshire, with her husband, the writer Howard Mansfield; their dog, Thurber; and eight hens.

One of the most widely read authors on anthropology and animals, wild and domestic, Elizabeth Marshall Thomas has observed dogs, cats, elephants, and human animals during her half-century-long career, all of which was inspired by her lengthy trips to Africa as a young woman. Her many books include *Dreaming of Lions*, *The Hidden Life of Dogs*, *The Social Lives of Dogs*, *The Tribe of Tiger*, *The Old Way*, and *The Hidden Life of Deer*. She lives in Peterborough, New Hampshire, with her dogs, Capek and Kafka; and cats, Coaly, Claude, and Shy.

the politics and practice of sustainable living

CHELSEA GREEN PUBLISHING

Chelsea Green Publishing sees books as tools for effecting cultural change and seeks to empower citizens to participate in reclaiming our global commons and become its impassioned stewards. If you enjoyed reading *Tamed and Untamed*, please consider these other great books related to nature and animals.

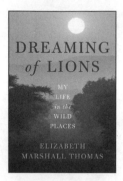

DREAMING OF LIONS
My Life in the Wild Places
ELIZABETH MARSHALL THOMAS
9781603586399
Paperback • $17.95

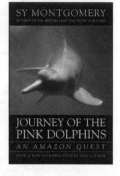

JOURNEY OF THE PINK DOLPHINS
An Amazon Quest
SY MONTGOMERY
9781603580601
Paperback • $24.95

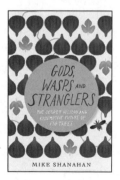

GODS, WASPS AND STRANGLERS
The Secret History and Redemptive Future of Fig Trees
MIKE SHANAHAN
9781603587143
Hardcover • $22.50

BEING SALMON, BEING HUMAN
Encountering the Wild in Us and Us in the Wild
MARTIN LEE MUELLER
9781603587457
Paperback • $25.00